Ancient Land, *Pastures New*

The Creation of a Small English Vineyard

Christopher Cooke

With many best wishes
and thanks

Christopher

First published in Great Britain in 2024 by West Berks Press

A catalogue record for this book is available from the British Library.

ISBN 978-1-3999-8815-5

Typeset in Adobe Garamond Pro 11pt

Designed by Chantal Bourgonje (chantalbourgonje.co.uk)

Printed and bound in Great Britain by Biddles Books Ltd

www.windingwoodvineyard.co.uk

For Leonie and Robert

'If you don't learn constantly, you don't grow and
you will wither. Too many people wither on the vine.
Sure, it gets a little harder as you get older,
but new experiences and new challenges keep it fresh.'
(Iris Apfel 1921-2024)

'The idea that in farming we are really exploiting
the land is quite correct. Indeed, we cannot help doing so.
With all that we send out into the world from our farms,
we are taking forces away from the earth – nay, even the air.
These forces must somehow be restored.'
(Rudolf Steiner 1861-1925)

'Wine is sunlight held together by water.'
(Galileo Galilei 1564-1642)

Further Reviews

'As judges we went over all the wines several times and we all felt Winding Wood Rosé was a shining beacon of the way forward for English pink. I think this is a VERY important category which we should be excelling at, but too many producers aren't yet quite achieving results. I have to say, two of Kent's top producers both told me how difficult it is to get pink exactly right - so - well done.

I particularly enjoyed reading Christopher Cooke's Ancient Land Pastures New because I knew that we had just awarded his Winding Wood Sparkling Rose 2020 the 2024 WineGB Trophy for Best of the Year. And his book allowed me to track the wine from being just a twinkle in Christopher's eye to being a fully-fledged, deliciously thrilling foaming pink beauty.

This book is at its best when you read it as a diary - chatty, informative, sometimes quite private, sometimes positively philosophical. And always honest. As Christopher says, "Life in the vineyard would not be the same without the regular existential crisis."

And so it proves as Christopher battles frosts and pests and the English weather and Brexit. There are no rose-tinted spectacles being worn here. I was particularly taken with a chapter on biodynamic vine growing which made more sense to me than far more intellectual and scientific treatises that I've read. And there are nine appendices which, for once, are actually worth perusing.'
Oz Clarke, OBE

'Having worked with several UK wine producers we understand the tenacity and focus required to plant and run a successful vineyard, let alone one farmed on organic and biodynamic principles. Christopher has created a gem at Winding Wood, constantly stretching himself and learning in the pursuit of excellence.'
Ben Llewelyn, Carte Blanche Wines

'For anyone keen on starting a vineyard this book is an interesting read, peppered with lots of good nuggets of information. Good colour photographs and line drawings and (as you would expect from an ex-publisher) the production values are excellent. Taken as a whole, it would make a great present for anyone thinking about starting a vineyard in GB.'
Stephen Skelton

Winding Wood Vineyard

Winding Wood Vineyard is planted on land at Orpenham Farm, in the parish of Kintbury, West Berkshire. It is an ancient piece of pasture. There has been a farmstead on this site for a thousand years or more. Mentioned in The Doomsday Book, Orpenham (the etymology is interesting with variations including Erpenham and Harpenham) is situated on a Roman road, Ermine Street, which ran from the area's capital of Silchester (Calleva Atrebatum) westerly towards Bath (Aqua Sulis).

The Cast List

Founding partner	Christopher Cooke
Founding partner	Robert Louth
Partner (life & working)	Leonie Cooke
Vineyard dog	Ludo (working cocker)
Agronomist	John Buchan
Viticulture consultant	David Morris
Wine maker	Daniel Ham (Offbeat)
Wine maker	Nicola Ham (Offbeat)
Winery dog	Jarvis (working cocker)
Viticulture	Paul Woodrow Hill (planting) Ed Mitcham (ongoing)
Vineyard help	Richard Pearce Martin Buckland Jane Buckland Polly Gibson Jayne Thorp
Vendage team	Too many stars to mention
Frost protection	Gaia Climate Solutions
Tractor driver	Tony Egerton
Website	Planet Design
Vineyard Guru	Hugo Stewart
Wine label design	Robert Dalrymple
Front cover photography	Charlie Stebbings

Contents

Preface

It seems a lifetime ago that we first planted vines in virgin soil. Yet, here we are in 2024 about to gird our loins once more and begin another season of grape growing. This will be our 10th vintage. It feels like a lifetime has passed. Happily, not a day goes by without my learning something interestingly new about wine making or viticulture. The body is more bowed, I have more callouses on my hands, but I hope the mind is still willing. I am certainly a wiser grape farmer.

A very good friend of mine once remarked, as we sat across table at a harvest lunch after the completion of an arduous morning of picking, that he could not believe how much blood, sweat and tears it entailed to produce clean, ripe grapes in England. He thought it as challenging, if not more so, than successfully raising children. I think he makes a good point, except that vines can be harder to discipline than offspring.

Would we have started out on this long, winding road to produce wine if we had known then what hurdles were in store for us, how absorbing, how maddening, the whole journey would be from bud to bottle? I do not know is the honest answer. I certainly did not envisage spending 10 out of every 12 months in the vineyard undertaking various critical chores to ensure healthy vines. Our vines really are special. They positively like our West Berkshire terroir. We have known each other since they were infants. Occasionally I will talk to them in an encouraging voice. While I have resisted giving each vine an individual name (bit tricky) they are without doubt the most pampered, spoilt vines in the country. I hasten to add that it has taken me a full 10 years to acquire the necessary skills to carry out winter pruning and vine trunk reduction with full confidence.

It has been a viticultural roller coaster with each season dramatically different from the previous one. Why? The unpredictable climate, of course, which in England can unfurl every tight roll of planning and preparation for the coming season. We are not alone as weather chaos increasingly occurs in every established wine region around the world.

There have been highs and lows, without question. The year 2020 was a low point when we suffered losses through a late frost in May. What made it worse was that the country was in Lockdown with many people at home sunbathing in faux jollity during an unusually hot spring. We felt utterly dejected. It followed that 2021 was a small harvest as a result of low budding in 2020.

Over the years, we have experienced the full gamut: plenty of good vintages, two great ones – in 2018 and 2022 – when the Gods looked down on us kindly; and a few miserable years, best forgotten, when we were decimated by either spring frost at the beginning of the season or mouldy, late summers. We have learned ruefully to take these bad years on the chin. This is the life of a vigneron. It helps enormously to be part of a supportive club of wine growers who will commiserate, sympathise, and celebrate with you in equal measure.

What started out as a project for two middle-aged, enterprising hobbyists with green fingers, grew into a passion, and then metamorphosed into something of an obsession which has taken over our lives for a decade and more. The ambition burns brightly always to grow better and better grapes, which in turn will produce finer and more complex wine.

Our travel plans each month of the year, bar November to January, are strictly determined by what needs to be done in the vineyard. It dictates when we can go away and when we decidedly cannot. It is no bad thing to be engaged and busy at my time in life in such a land-based enterprise, exposed to the vagaries of the weather.

As an inveterate scribbler, the creation and development of a vineyard in 'my back yard' provided perfect material for my pen. I therefore kept a diary recording the seasonal rhythms of the vineyard and the exhausting work we undertook to grow fruit in a cool climate, sometimes on the edge of what was possible. Part of that diary has appeared as a monthly blog on the Winding Wood vineyard website that one customer described as rather like reading a script of an Archers' episode, minus the human intrigue.

Plenty has been written on wine and vineyards, much of it engaging. The English wine scene is covered on a regular basis by the media largely in glowing, if sometimes, hyperbolic terms. For many outside the industry, the thought of owning a vineyard, growing grapes, enjoying bucolic bliss, is one of pure and unadulterated romance. Wish bone in place of back bone, as the expression goes. Try working outside wearing four layers of clothes during the cold, winter months with numb hands!

As a counterbalance to this rose-tinted view of vineyard ownership, I felt that a 'behind the scenes' account of how that fine bottle of English wine eventually sees the light of day on the table of the restaurant diner, would be both entertaining and insightful, not only to wine-loving readers but also to potential vignerons. What follows is, I hope, a warm draught of reality.

Introduction

My Wine Odyssey

My first memory of being interested in the drinks industry was at the tender age of six. I had cleared my bedroom of all furniture apart from a bar created out of a plywood sheet placed on the top of my chest of drawers. My family were all invited, at the appointed hour, to 'purchase' a variety of soft drinks which I dispensed into water glasses borrowed from my mother's dining room sideboard. I rather took to being a publican of sorts.

There then followed a gap of fifty years before I was financially and temperamentally in a position to make my own alcohol.

In my early teens, father's drinks cabinet was the first port of call although that became more difficult as we children became keen on gin and our father became suspicious that the bottle was emptying rather dramatically. He then began to mark the level after pouring himself a drink. On his return from the office each day he would treat himself to a gin and martini whilst listening to the stock market results on the radio. If the news was gloomy, he would often pour himself another.

Thanks to a good stock pick with Farnell Electricals (a tip from stockbroker, Tim Doyle) we began to take family summer holidays in Normandy. Father landed in the early hours of D-Day in 1944 at Bernières-sur-Mer (Juno Beach) and disgorged young Canadian troops onto the beaches, the poor sods. He survived, so there was a good reason to return. We would stay at a lovely hotel on the quay and enjoy meals accompanied by small tasting glasses of white and red wine. The food was in a higher league than the plumbing.

Visits to my great aunt Beryl and great uncle Christopher for lunch in bucolic Upper Basildon, Pangbourne were usually accompanied by watered down French red wine (vin ordinaire) for us young ones and home-made mints which were glorious. As we left, Tom (Tired Old Man) as he was referred to, would press a florin into each of our palms while, for amusement's sake (as I thought at the time), he would drop a coin down the front of my mother's dress. She would wriggle at the steering wheel all the way home. If we were lucky, Tom would hand us half-a-crown.

At university I eschewed joining a wine society, (if one existed which I very much doubt) as I was far too anti-establishment. Instead, I gained a liking for Yorkshire beers and real ale. In London, after graduating, the craze after work was to drink as much Beaujolais nouveau as one could in a wine bar and then try and mop up the effects with a curry. What a clever invention – those French were brilliant marketeers, but the wine was quite, quite dreadful.

In my late twenties, New World wine became the rage – helped along by Oz Clarke who became the wine correspondent for *The Sunday Times* – with intoxicating reds from Australia and those over-oaked Chardonnays. My first plunge into decent wine was through a wine outfit who used to come to both office and home.

Over several years we bought plenty of lovely Burgundies, such as Gevrey-Chambertin, which we 'laid down' in our wine cooler without much experience of how to maintain a constant temperature. Few Notting Hill houses had underground cellars. Many fine wines we left too long. Moving house regularly did not help the wine. Our dinner parties became legendary for uncorking bottles of Burgundy, only to find them well past their best and then pouring them straight down the sink.

We were thoroughly spoilt at our wedding at which my father-in-law, Maurice Lawson, most generously served Roederer Cristal throughout the lunch party for 30 guests. I was too preoccupied to sup much but my uncle Stuart, who knew a thing or two about champagne, was in his element and kept demanding the waiter re-fill his glass throughout the meal. He said it was by far the best wedding he had ever attended.

Robert Louth and I became good friends when I joined the shooting syndicate at Kintbury Holt in my forties. At that time, I was a very light wine drinker. Robert used to tease me for being a 'flat cap' wearing beer drinker. He had been 'in training' since medical school days. It was certainly Robert and another local friend, Jeremy Hawkins, who introduced me to the delights of Rhône wine, and I joined The Wine Society on the strength of their suggestion. When the Society's En Primeur offer was published each spring, we three would all mark up the thick catalogue and compare notes. Building a cellar was an inevitable consequence of buying to 'lay down'. I had a lot of catching up to do with these two wine aficionados, but I made amends by coming out of the blocks fast. I followed Robert's suggestion and based my cellar around Rhône, Bordeaux, Burgundy (white), Sauternes, and a smattering of Spanish red.

By the time we began to produce Winding Wood sparkling, it became clear that a second, temperature-controlled cellar in the barn was required, if only to save the cost of storage at the winery. Capacity for 5,000 bottles seemed quite modest. Thank goodness, common sense prevailed, and we did not attempt to excavate a huge cellar below the ground of the granary that would have needed a lift system to bring the wine pallets up and down.

Therefore in 2012, a wine summit was called with our respective wives to sound them out on the project. Frankie Louth was very supportive, not least because it meant Robert would be fully occupied when not sailing – rather than hanging around the house getting in her way. Leonie, my wife, pronounced that she had married me for life but not for lunch. We had the green light to proceed.

Chapter One

What, A Vineyard in England?

Throwaway remarks can have unforeseen consequences. As I sat with my great friend, Robert, having a sundowner one summer in 2011, we gazed upon a flock of Jacobs sheep nibbling away happily in our paddocks. Having been a farmer in an earlier life, with many more herbivores under his care, Robert sympathised when I remarked ruefully that they were a lot of trouble for the return.

'Then why don't you plant a vineyard instead – and then you can drink the results,' he chortled. This remark planted a seed in my mind and two years later, after hours of research and visits to many vineyards, the fields in front of our farmhouse were ploughed up, sub soiled (oh, the rubble we found) and a few thousand vines were hand planted into virgin soil. Oh, and Robert was roped in as my vineyard partner.

My retirement plan then went horribly wrong. Roll forward twelve years and if you happened to pass by the vineyard at the end of any October you would witness the two of us plus a troupe of green-fingered volunteers patiently moving down each row of vines, neatly snipping off bunches of grapes into awaiting crates.

Of course, English still wine has been around for years – in varying levels of quality and certainly nothing to match a good white Burgundy – but we are part of a new wave of English wine growers who are making English fizz by growing the classic champagne varietals and then using the traditional methods of two fermentations. In southern England we share much of the geology with Champagne in northern France; and terroir-wise, with chalk and flint, the soil is ideal for growing Pinot Noir and Chardonnay grapes. Then add the perceptible change in the UK climate and you have the perfect (potentially) growing conditions.

You need patience coupled with resilience and it is a good idea to start out on such an enterprise before you enter your 60s because it is a long, hard road – 5 years at least – before you get to taste your first bottle of fizz. We released our first vintage, the 2015 Winding Wood sparkling, in the latter half of 2018 with some trepidation and excitement. We could not wait to pop the first cork. To date, we have released five vintages with another four more in the pipeline: the 2020, 2021, 2022 in various stages maturing on their lees (bottles on their sides in contact with the dead yeast); and the 2023 wine is fermenting in oak barrels – all at Offbeat Winery, nr Salisbury, where we are fortunate to have Daniel and Nicola Ham looking after our precious wine.

Along this path of wine adventure, we have won a few wine gongs in national competitions which is immensely satisfying. Our second release, the 2016 classic cuvée, won a clutch of medals including a gold from WineGB. And our 2018 wines have harvested five medals. In 2024, two major trophies for our Rosé 2020. Not bad for a couple of oldies from a standing start.

Friends often remark that, after this total immersion in grape growing, we must have learned all there is to know about viticulture. If only. We normally retort with the old saying: 'The more one learns the less one tends to know about a subject.' That is certainly the case with viticulture and wine making. In ten years, we have just scratched the surface of what one needs to learn to be an expert.

If we had known then the trials, tribulations and occasional tears of frustration involved in the process, I wonder if we would ever have set out on the journey to produce the very best fizz we could from our land. Both of us had hung up our professional 'boots': Robert, his dentist's drill and yours truly, the monotony of publishing deadlines. After 40 years of hard work, we were looking forward to taking things a little easier. Not a chance, as it turned out. The new hobby has become all-consuming with 250 days a year spent in the vineyard. November and December are the only months of viticulture inactivity.

Along the way we have met some delightful and passionate wine growers in England, shared valuable information and the odd calamity, and joined an exclusive band of vineyard owners who firmly believe England can produce the very best sparkling wine to match that from Champagne. The major awards from blind tastings speak for themselves. It is, without doubt, the best club I have ever joined.

Small Can Be Beautiful

We have deliberately kept things on a small scale and resisted expansion, with just 2,500 vines in two small fields of Pinot Noir and Chardonnay. The goal is low yield and high quality. Every vine is subjected to continuous scrutiny and exacting canopy management throughout the growing season to ensure the grape bunches are healthy and free of disease. As our wine maker continues to impress upon us, he cannot make good wine out of poor grapes. For sure, with every year of production we become more in tune with our vines, knowing how to control their desire to run amok and shoot for the sky. That is the pleasure. Almost all of the vineyard work is done by us, come rain or shine, with a little bit of help, in the early stages, from green-fingered volunteers. But now a trained gang of Romanians under the direction of Doru help us. He never ceases to surprise me with his immense knowledge. At the end of the day the back may be sore but there is nothing more satisfying than looking out at our vineyard and seeing the results of our toil.

Never wander into the vineyard without a pair of secateurs

When, back in 2013, we planted our vines, we committed ourselves to the production of sparkling wine made in the traditional method of two fermentations, reflecting the unique terroir and character of our small Berkshire vineyard.

Our soil is ideal for English fizz as we are planted on free-draining loam over flint and chalk on a gentle east-south-east facing slope– good for catching the morning sun which dries the vine leaves. Each vintage we make consists of only grapes from that year's harvest (with the occasional small exception). Grapes painstakingly grown, lovingly nurtured and then meticulously picked by hand.

One of the advantages of being small scale is that one can experiment with the latest technologies without breaking the bank. One of the major worries of growing grapes in a cool climate is the likelihood of spring frosts that can devastate young vine buds. Having seen heated wires protecting the vulnerable buds on a few experimental rows at Ridgeview in Sussex, we decided to install wires throughout the vineyard. When the temperature drops to below zero, a thermostat automatically switches on the system transmitting 20° C of heat through the vine cane creating a halo of protective warmth against the frosty air. We have tested it and found it good to minus 4 ° C. We believe we are the first in England to install this revolutionary system throughout the vineyard. It has saved our bacon.

At the outset our wine was made for us by Emma Rice at Hattingley Valley. In 2021 we moved on to Daniel Ham of Offbeat Wines. Daniel is a marine biologist by training. Having re-qualified at Plumpton College, he started his wine-making career at Ridgeview under the legendary Mike Roberts, then became head wine maker at Langhams in Dorset before embarking on his own at Offbeat. He makes his own biodynamic wines and, additionally, wine for a number of small producers like us. Increasingly I am convinced by Daniel's view that the best wine is made in the vineyard not in the winery.

A Very Brief History of English Wine

It is pure romantic conjecture to think that the Romans planted vines in England.There is limited evidence albeit times were most probably warmer. If they planted any vines, it certainly would not have been the aromatic Bacchus grape variety – even if Bacchus conveniently was the Roman god of wine and pleasure. For the very good reason that this variety was only developed in the late 20th century! More credible is the notion, given their legendary logistical skills, that those Romans based in Britain imported their wine from the hot parts of their empire.

Fast forward to the 15th century. The Dissolution of the Monasteries, on the orders of Henry VIII, had the devastating effect of ejecting monks from their living, thereby causing a huge loss in wine making skills throughout the land.

Fast forward again five centuries to the 20th century. After World War II, a few returning soldiers, maybe seeking solace after their wartime experiences, planted German varieties – including the famous Sir Guy Salisbury-Jones who planted vines in 1952 at Hambledon Vineyard (the village where cricket was first played circa 1750). Not surprisingly, they found it difficult to ripen the fruit. The wines were light and aromatic but a bit of a joke. and gained a richly deserved reputation as being not very nice.

If Act One in the English wine world gave us plenty of amateur enthusiasts giving it a go, Act Two saw the dramatic entry on stage, in the late 1980s, of sparkling wine using the traditional champagne varieties – with Nyetimber in the title role. Stuart and Sandy Moss from North America discovered the soils in West Sussex to be good for growing the classic champagne grape varieties of Chardonnay, Pinot Noir and Pinot Meunier. It had the same geology as in Champagne. Their punt paid off and Nyetimber continues to grow in quality and world-renown under the present owner. So, the revolution began.

The English Wine Revolution

Unlike in other European countries like France, Spain, and Italy, where the wine industry started on a small scale with peasant farmers and then evolved, we have witnessed an extraordinary revolution in England over the last 10 –15 years with big money from wealthy investors being thrown into planting large vineyards – the great majority to be found in Kent, Sussex, and Hampshire – many with slick cellar door offers. Many new owners have no agricultural or fruit farming experience. Perhaps they own a grouse moor, so why not add a grape moor.

According to Susie Barrie, Master of Wine (MW), we can now think about Act Three: the coming age of seriously good English still wines. A recent report written by a viti-climatologist, Alastair Nesbitt, predicts that climate change will benefit our still wine growing in parts of England. Essex may become the UK equivalent of the Côte d'Or in Burgundy. I am most suspicious of forecasts (just remember those modellers of doom from the Pandemic) yet, having seen the fruit coming into OffBeat from Essex in 2022 with a potential alcohol of 14% – much, much too high for Chardonnay in my opinion – then one does wonder what the next few years will bring.

Realistically, English still wine is still being held back by not enough sunshine. The financial margins are not there. You need to be very patriotic to pay upwards of £25 for an English Pinot or £30+ for a Chardonnay produced in Essex or Hampshire. Plenty of fruit for sure but not much complexity.

'If you want to make a small fortune out of owning a UK vineyard… then best to start with a large one.'

The investment being made is huge, most of it in sparkling wine plantings, yet returns remain a long way off. Some estate owners will have invested north of £10m. Just take the yields in viticulture versus those in agriculture. If you take the 2022 grape harvest as a good average at 3 tonnes/acre, then look at the top and bottom of the range: years like the exceptional 2018 of 150% versus bad years of 50% or 70%. In the terribly wet year of 2012, no wine at all was produced. Carnage to the balance sheets. Whilst in agriculture, the swings can be +/- 10 to 15%, in viticulture it is a financial knife edge where unpredictable weather can wreak havoc with plans.

The growth curve is impressive in the UK, there is no denying it. We have grown from vine plantings in 2000 of 857 hectares (1 ha = 2.4 acres) to 3,781 in 2021. There is a disparity in size. As at 2023 three quarters of the 1000 vineyards in England and Wales are just 2 hectares or below – therefore small commercial producers – whilst the remainder are very large estates. Quite dramatic growth when one considers that there were only around 400 vineyards when we planted at Winding Wood in 2013.

At the top of the league table, you have large wine producers such as Nyetimber with sales in 2022 of 1m bottles pa, Chapel Down with 1.7m, and Hattingley Valley with 0.5m (impressively 40% exported). They have found export markets in USA, Scandinavia, and Japan but export sales for our industry as a whole are still only 4%. Compare this to a medium-sized champagne producer like Lanson who sell 4m bottles pa – 75% to export markets –and you realise that we are still minnows compared to our friends across The

Channel. However, the trend is good. In 2022, for the first time, UK annual sales matched production, that is 9m bottles. Up to that point, production far exceeded sales – take the bonanza harvest of 2018 when 13.1m bottles were produced against only 3.3m bottles sold.

Again, let us put things in perspective. The UK region in terms of annual sales is only the size of Oregon (or indeed Lebanon for that matter). However, as the newest of the New World regions, the industry is growing at speed, with even the French investing in England. Tattinger owns Domaine Evremond in Kent with 40 hectares. It is expected that they will release in 2024. Louis Pommery has planted in Hampshire with their wine made for them by Hattingley Valley. Freixenet Copestick has bought Bolney Estate; and the Jackson Family, a Californian producer, has purchased land in Essex with the aim of planting still varieties. The attractions? Climate change for sure and the cost of land. A hectare (2.4 acres) in highly-sought-after regions in France would cost €1m euros versus £30k for a prime acre (£72k/hectare) in England. You can see why French producers would want to invest across La Manche. Cheap land for sure but also the only commercial viable response to domestic competition.

As a cool climate region, English grapes would be considered under-ripe in other parts of the world where they produce 'quality' still wine. However, our conditions are ideal for producing sparkling wine. Remember sparkling grapes will be picked at 10 –11% potential alcohol. Any higher would be unwelcome as the second fermentation will add another will add another 1-1.5% of alcohol.

In the hot summer of 2022 (it hit 40°C in July), we were slightly concerned at Winding Wood that our grapes were becoming too ripe, and potentially losing levels of acidity essential for making good sparkling wine. There are, it is reported, parts of Champagne that are getting too hot. In the scorching year of 2018, some growers and winemakers were very worried that the grapes were becoming too ripe with plummeting acid levels. Many winemakers interrupted malolactic fermentations (where malic acids converts to the softer lactic) for good reason.

Within just five miles of our vineyard, in Boxford to the east and

Shalbourne to the west, there are now two large vineyards planted by wealthy businessmen, both canny but with radically different visions, who are prepared to join the wave with seemingly unlimited money. Alder Ridge, just outside Hungerford, originally planted by Cobbs Farm Shop, has recently been sold to an operation, High Weald, based in Sussex. They have expanded the existing 5 acres by planting another 25 acres of Chardonnay on what were fields of asparagus and soft fruit plantings. This is their first experience of terroir brimming with flint and chalk. Their equipment may take a battering.

Whether, when the high volume of wine from these extensive new plantings comes on stream, it will have an adverse effect on the market, time will tell. Some doom merchants predicted that the frenzy of over-planting would produce a huge mountain of surplus wine, yet to date this does not seem to be playing out. The UK is a huge market for wine with English production very small by world standards, so if the UK producers can capture a small percentage of that domestic consumption (plus exports) they should be fine for the foreseeable future. Some argue that producers of 'quality' wine will be viable for even longer, maybe for ever! Other wine commentators have forecast a consolidation in the industry where medium-sized producers are caught in the middle, being neither boutique nor large. No doubt, unless they have established good markets for their wine, they will get squeezed on price and gobbled up by larger concerns with superior financial muscle. Watch this space.

Chapter Two

Planning and Planting

The spring of 2013 saw us digging bore holes, choosing a viticultural agronomist, researching grape varieties and visiting as many vineyards as we could to build up knowledge. We had the foresight to secure our winemaker even before we had a vine in the ground. Given that the English and Wales wine industry was still half the size of what it is today – maybe 400 vineyards at most – there was at that time a dearth of places where one could have wine made well. Our forward thinking was very fortuitous in hindsight. To build a winery here at Orpenham was out of the question: a mammoth and expensive undertaking and only viable if one was processing north of 40 tonnes of grapes per annum. This, we were definitely not planning to do.

John Buchan, our agronomist, recommended we approach Emma Rice, head winemaker at Hattingley Valley, who had won two awards for UK Winemaker of the Year. But what to plant, we asked her? German or French varieties? Emma told us in no uncertain terms that while we could grow German still wine varieties, as had been planted in England for decades, she would be unable to make anything decent with them. That was the spur we needed towards a decision: we would grow champagne varieties and Hattingley Valley would be our contract winery for producing our English sparkling wine.

Robert's fields at Berries Farm were clay-bound and totally unsuitable, whereas the front paddocks at Orpenham Farm had potential. They had been grazed for centuries but no ploughing or planting as far as we knew – we believe there has been a farmstead on this site since the 11th century (mentioned in The Doomsday Book). The elevation was high at 130 metres above sea level – thus cold – and it was east south-east facing. The ground however was ideal with loam followed by flint followed by chalk — perfect free draining soil for planting champagne varieties. One of the fields was pH zero perfect, the other required balancing over time. We did not know at the time that the site was a frost pocket with nowhere for the cold air to roll off, but we had to work with what we had. Furthermore, no commuting to work for me.

Choosing and ordering vines is no easy task. We selected a company called Vineworks to order our vines. It's a tricky business, particularly as we were so green in these matters. First of all, we had to select the rootstock and the clones. Following advice and looking at our soil, it was decided to plump for S04 as the 'American' rootstock for both the Chardonnay and Pinot Noir; and then clone 95 for the Chardonnay vines and 375 & 395 for the Pinot. These are both Burgundy clones. The vine clones are grafted onto the rootstock and held in place by red wax. We had wanted to plant the trinity of champagne varieties but there was no room to have Pinot Meunier - the third and minor grape of the three - in my humble opinion, in a decent enough quantity to press at Hattingley Valley. Their presses required in excess of 1.5 tonnes of grapes in order to obtain a good pressing. We know differently these days having moved to a smaller and more flexible winery with basket presses capable to taking just 200 kilos. Anyway, the die was cast and we plumped for just two varieties.

Soil preparation

The next task was to kill the grass and then plough the two fields; followed by sub-soiling and rolling prior to planting. This was no mean feat and one I let Robert organise. We had two tractors lent to us by Andy Stevens, a local farmer, one of which drove all the way over from Lechlade. They did a beautiful job in such a small space of 1.5 acres. For sure, it made a change for Matt and his fellow tractor driver from ploughing a 50-acre field. Along the way the rubble and large sarsens were dug up and put to one side.

Planting

We started by planting the Chardonnay vines. Come May (after spring frosts had passed) the vines arrived, shortly followed by a gang of Romanians to plant them. As you can see from the photographs, there was no satellite equipment in play as we marked out the fields into. rows. And no planting machinery — we were too small for that. All the vines were planted by hand with a bit of help from petrol-driven augurs. These hit large flints on many occasions which may explain why some of the vines are not strictly 1 metre apart.

One major departure we made in planning the vineyard was to decide on narrower row widths of 1.8m rather than the traditional 2.4m. This has implications for the choice of tractor and equipment, in our case a mini tractor; the flip side being that our density per acre was higher than most vineyards. That potentially equalled more wine production! The norm is 1,500 vines per acre yet we managed nearer 2,000.

Each vine has a 1metre metal tutor for support and a rabbit guard to protect it from nibbling. The tutors are eventually clipped to the fruiting wire once the trellising has been installed. The posting and trellis wiring for the Chardonnay field was installed by us for the simple reason that it was (a) much cheaper and (b) Robert loved the challenge. As an ex-farmer, he thought this was straightforward. It was back-breaking work piling posts 3 feet through very unyielding flint where the post basher could not get close, we had to prepare holes the intermediate posts by hand. The end posts were tricky — they would take all the eventual weight of the canopy — as they needed to be put in at an exact angle and then anchored underground.

By the end of the summer the vines, thanks to plenty of rainfall, were were beginning to protude above the rabbit guards and looking very perky. This was just the beginning of the odyssey.

Our order was placed too late to buy the Pinot Noir in time for planting in 2013. There was plenty of grinding of teeth followed by the dismissal of the contractor. In 2014, we used another contractor who not only sourced the vines but came, planted and trellised the plot. The best outcome for us was that we did not have to expend sweat and blisters getting the posts in ourselves.

When it came to installing the posts and wire in the Pinot the following year, we left it to Paul Woodrow Hill and his gang to bring in proper equipment for post bashing. Within 20 minutes the operator had managed to bring down the telephone wires which straddle the field. No phones for the neighbouring Hamblins!

It is always a dangerous business ordering large amounts of vineyard materials as one has no idea what size of pantechnicon will turn up. They see 'farm' on the address and assume there will be a huge yard with a forklift truck at the ready. No doubt the driver will not know how narrow the lanes are around here, nor will he speak English. That turned out to be the case on both fronts when a 30-tonne truck appeared without warning carrying our metal posts and reels of wire plus equipment for other vineyards including a tractor. The driver could only grunt and use sign language. Out came the Landie and trailer and slowly we unloaded from the roadside. There was a lot of swearing under the breath. The driver took out a few metres of hedging getting the monster back on the road to the A4. Never again (until the next time).

2010. Land before we planted. Sheep nibbling

2013 April. Two blue tractors meet across a field

2013 October. Robert Tarzan supervises end post installation

2014 May. Rabbits beware!

2014 June. Is it okay if I bring down the telephone wires?

2014 September. Chardonnay vines with barn in the background

2015. Christopher poses

2015. Robert poses

2016 April. Robert aboard the fertiliser spreader

2015 May. Robert manure spreading

Chapter Three

A Typical Year in the Vineyard

Friends often ask what I do during the summer before the harvest. I roll my eyes and emit a 'tut, tut' sound. Clearly, they think that the vines just look after themselves, a bit like a raspberry cane, producing fruit to order. Oh, I wish. The vine is a woodland plant and, given a free rein, would go rampant and shoot for the sky.

Here follows a record of a calendar year in the vineyard. As will become clear, there are precious few dormant months.

January

Feet up for three weeks. Towards the end of January we don coats and start to pull last year's cane and shoots off the wire. This creates enormous piles of cuttings which are fed into a shredder and then composted. Frost and snow are the norm at some stage in either January or February. Best not to try and do anything outside on days like this one for fear of frostbite.

February

Sheep will graze. We borrow ewes from local farmer, Trevor Gore. They are not only excellent herbivores but provide good manure for the ground. That is, assuming they do not chew the vines.

March

Winter pruning kicks off by cutting away last year's growth from the wires, then selecting this year's cane and spur (the cane for next year). We use the single guyot pruning system (see appendix 1). Ed Mitcham demonstrates his technique while Robert has electric pruning shears to ease the arthritis in his hands.

April

Life begins to move. Bud burst and first leaf out. This is when the buds are at their most vulnerable to spring frosts ....and encroaching nibbling deer. Clear skies at night with an abundance of stars mean trouble. In the early days we used to light bougies on frosty nights with a pre-historic telephone with alarm in a box which called us when the temperature dropped during the night. Later, we moved to a heated wire system activated by a thermostat which meant we slept soundly.

May

Lots of leaf out. Spring frosts are still a possibility until mid-month. Plenty to do removing double buds on the cane, bud rubbing away the unwanted buds on the crown and trunk, adjusting cane ties, sorting out the canopy wires. Keeping the grass down is a major task and it grows as one looks at it. Between the vines we use a strimmer with a guard to protect the trunk.

June / July

Temperatures rise and flowering begins, followed by 'caps off' and bunch formation. This is a critical month. Poor flowering will knock any season back by weeks. The flowers are very delicate and easy to miss on a stroll through the rows.

July

This month the canopy really motors and we have to keep it under control. Removal of laterals in the fruiting zone. Robert stops to chat with Tony who has his tractor with an air-assisted sprayer attached all ready to go.

August

This shows pea sized berries in the Pinot field. We strip leaves from around the bunches to maximize sun penetration. The earlier this is done once flowering is largely complete, the better, leaving enough leaf above the bunch for good photosynthesis.

September

Full and lush. This shows an Indian summer in 2020 offering great promise for a ripe harvest. Green harvesting begins where we take off bunches which may not ripen in time, to encourage the remainder to ripen. We continue to monitor volume

estimates for the winery so they can plan. No surprises wanted at this late stage, such as not having enough tank capacity.

Early October

Veraison starts this month (or can be earlier depending on the weather). Bunches swell and change colour. We are now in the run up to harvest. Keep disease away. Green harvest every couple of days.

Late October

The Pinot Noir turn a dark hue of purple – now we know they are ready for picking. Get out the picking crates to clean and get ready. Sort the trailers. Email the pickers list to get notice of their availability.

October Harvest

Above, the dream picking team for the 2018 Pinot Noir. A fabulous summer and bumper crop. Everyone was on best picking form. A long lunch followed afterwards – a blessed relief.

October - The Winery
Our Pinot grapes are shown piled in the Coquard basket press ready for a gentle squeeze. One needs to be fit and strong to work this press, with lots of forking to be done between pressings.

November
The end of the season draws nigh. The vine leaves turn to yellow and then drop. Trips to the winery to inspect the wine.

December
It is now cold. The vines want to go into hibernation. They have worked hard all summer to produce the fruit and have earned a rest.

Chapter Four

Diaries, The Early Years

(2015-2017)

The following extracts are taken from the vineyard diaries for 2015, 2016, and 2017. In retrospect, these diaries are revealing in those first few years for how much we had to learn about tending our young vines. We were positively green!

The Year of Our First Harvest

February 2015

Who says there is nothing to do to the vineyard in February? Tom Bartlett, of Stopham Vineyard in West Sussex, has kindly agreed to come up and teach us how to lay down the Chardonnay canes in anticipation of our first harvest this summer. We have decided to go with the single guyot method of vine trellising which means that the shoot will be bent over in one direction (as opposed to a double guyot) and the shoots will grow vertically from the horizontal cane.

When we first met Tom, I could not place his slight accent and thought, with his viticulture background, he was an Australian; but he put me right immediately: 'I am from Essex'!

Cane laying. At first, we are both a little nervous of breaking the vine as we bend it over onto the fruiting wire but Tom strides down the row in confident fashion and demonstrates those canes which are thick enough in girth to bend and how to bend them slowly over and attach with paper ties. Selection is key because there is no uniformity with a Lilliputian plant growing next door to a giant (Brobdinagians, I guess?). Is this due to poor planting, we ask ourselves or just what lies beneath each vine?

First, off come the rabbit guards, next on go the rubber bands to support the main shoot against the metal tutor, and then the cane is tied down with twine to allow 4 buds to produce (our max first year allowance). Some of the vines whose stems are not yet thick enough to bend over will still produce a few bunches of grapes, says Tom, but we need to be careful not to the jump the gun and stress the vine into producing too much fruit before it is ready. I wish we had started this game a few years ago as Mother Nature takes its time.

'How many tons will we get this year, Tom?', we ask tentatively. There is a lot of head scratching and the phone's calculator is brought to bear. It is not quite clear as yet, but he thinks we might get 300-400 kilos.

26 February. 'Fail to prepare or prepare to fail'–an old adage my old school master used to employ as exams loomed, and of course he was right. Last year we got a nasty shock in April and also May – after the bud burst - when we were hit by several clear nights when the temperature went down to -2ºC. This killed off many young buds. 2nd May was a legendary -3ºC and caused havoc up and down the country. It was also the cause of us getting no cherries that year as all the blossom was hit badly. Last year we installed thermometers at various parts of the vineyard in a bid to discover where the frost pockets lay, and this task was repeated. The area down by the hedge next to the road was a chilly area — a frost pocket as it is called – and also the area on the hedge line of the Pinot seemed exposed.

Robert and I have been investigating systems for frost protection: from the very expensive frost busters which run on Calor gas and throw out hot air to dispel the frost, to the old-fashioned French method of lighting candles (bougies). There have been experiments with heated wires, similar to those installed on oil pipelines in Siberia, which are attached to the fruiting wire. The installation cost is hugely expensive. The idea of flicking on a switch from beside the bed and witnessing the temperature in the vineyard rise by 5 degrees in a matter of minutes is very appealing… but sadly the jury is still out as to whether it actually works.

Most vineyards in England (apart from those on the coast) suffer at some time or another from spring frost due to the fact that we are all growing vines at the extremity of what is possible given our climate; whilst in France, notably Burgundy, the enemy is hail.

There is only one French manufacturer of bougies, it would appear, and they are located inconveniently somewhere south of Lyon. Bougies come in 10 litre tin cans and are composed of oil wax and a rudimentary cardboard wick. Not to be mistaken for the other type of bougies, which is a 'slender, flexible, cylindrical instrument that is inserted into a bodily canal, such as the urethra, to dilate, examine, or medicate'! I think not.

After many calculations we decide to buy 800 bougies costing €5,000 (a lotta lolly), hoping this would last us for several years. The cost of transport is high as they are transported on pallets all the way from south of Lyon. Theo, my son, who is studying in Aix-en-Provence this year, was enlisted to discuss the logistics in his best French with the supplier and to ensure that the delivery was tracked on its journey so that we had a forklift at the ready when the lorry arrived at Kintbury Holt Farm nearby. Despite the best laid plans of mice and men I had a call on the day in question from the delivery company to say that the driver was parked up in the farmyard and where were we? Even in this day and age of mobile phones, why does one bother! It turned out that the driver had arrived in Kintbury with 2 minutes to spare on his tachometer so, after taking off the load, asked to bed down in the yard for his required 10 hours rest.

March 2015

Time to apply manure. We go to work moving manure from the compost heap to the vineyard. Capability Louth devises a system with tractor and trailer chugging slowly down each row with a man and shovel behind. It soon becomes clear that there is nothing quicker than two wheelbarrows and two shovels. We dress the base of each 'runt' vine with a shovel-full of well-rotted compost. Our neighbours, The Quintavalles have just taken delivery of 10 tonnes of Progrow from the local recycling centre and, with permission, we help ourselves greedily to some of the steaming pile. It is easy to handle and no weeds.

14 March. Apply the first spray of the season by knap sack of headland sulphur on the advice of John Buchan who says it will help build up protection against disease in the vine. Our other advisers think we are nuts! The opinionated world of viticulture!

18 March. After much desk research, and after threatening to make one himself, Robert locates a suitable air-assisted sprayer that will fit behind our tractor from a supplier in Evesham.

April 2015

Robert persuades Andy Stevens, a local farmer whom we shoot with, to lend him a family heirloom — an ancient fiddle for broadcasting seed which looks more like a bagpipe. It is with glee that he fills its bag and winnows down each row spreading boron with metronomic precision. Within an hour the wooden handle is broken but CC comes to the rescue by digging up some copper tubing and drilling a couple of holes in each end for the string to go through. Job done.

7 April. It is key for the viability of the vines to keep the grass ride under control so that the grass and the vine do not compete for nutrients. We want each vine to be free from any weeds and grass. I volunteer to drive the tractor with spray tank attached with Robert in charge of applying Roundup by way of a lance. It is hard to keep the orange bucking bronco at a slow walking speed (we have no rev counter), so Robert is almost skipping down the row behind me and by the end of it he is sweating profusely. 'Time for a beer', he cries.

When we installed the posts and trellis in the Chardonnay field in 2013, we had great trouble getting the anchors for the end posts at the end of each row inserted correctly (they will take the huge weight of the canopy) with the result that the end posts were beginning to move towards the vertical position. New plan – extra anchors for each post – which meant creating 54 holes with a metal post basher to a depth of 3 feet. Luckily Richard 'Turbo' Pearce was on hand to help with the manual duties but even he wilted towards the end.

Off to the Thames and Chiltern Vineyard Association AGM at Stanlake Park near Twyford, Berks where the new owners are kindly hosting the event. Meet the delightful owner of Dropmore Wines, an ex-marketing director of American Express, who could not have chosen a better name for a vineyard if he had tried, and we exchange stories on frost prevention. The camp is definitely divided into those that hire help to manage their vines and those that get down and dirty their hands. Fascinating mix of folk as usual. I sit next to a Master of Wine who is charming but whose hands begin to tremble a little as the clock marches past midday with no sign of refreshment! Robert has a good conversation with the owner of Stanlake Park about what one can and cannot put on the back label of a wine bottle.

23 April. Our sprayer arrives with Edward from Weavings Agricultural Machinery who will fit the thing and make sure it attaches correctly to the Kubota. Tony Egerton (our air assisted spray operator) and Robert spend the morning spraying water on the vines to test the nozzles. Too many levers and switches for my liking and I shall not be too hasty in coming forward to volunteer spraying duties. I am qualified for knap sack spraying but little else at this stage.

24 April. The weather report looked ominous and a frosty night it was. I was up at 1am to light some 200 bougies that we had spaced evenly through both fields. Using a blowtorch in one hand and a watering can with a mix of petrol and diesel in the other, and of course aided by my headlamp, I started my tour of torching each bougie. Under a starry sky, I staggered from one bougie to another, kicking off the lid with my boot, pouring in a quick slurp of fuel over the top of the wax and firing up the wick to catch fire with a 'boom'. Job done by 2.30 am. Luckily the local fire service did not come to visit as I think they would have been rather bemused by seeing two fields alight with a red glow as if UFOs had just landed. In the morning we extinguished the bougies by putting their tin lids back on.

May 2015

John Buchan, our agronomist, makes his fortnightly visit. He is based in Shropshire but spends his time wending his way up and down England visiting vineyards in every county. As a result, he has a very good grasp of what is happening on the ground: vignerons are like farmers in that they tend to moan all the time, as growing conditions are never quite what they want. It is either too cold or too wet. John is 'a cup half full' man with broad shoulders, so a visit from him always peps us up.

He is extremely knowledgeable, advising as he does some of the biggest vineyards in the UK... and some of the wackiest. I am not quite sure what he makes of the Louth/Cooke team as it is both a mixture of blue and white collar but at least he finds us eager to learn every nuance of viticulture. Robert lets him get away with nothing and grills him for every piece of information.

His advice this time is that we should install a top wire in the trellising to prevent the vine tips from being damaged by flapping in high winds. Luckily, we have invested in a spinning jenny which means we can uncoil rolls of 2mm wire over 100 metres without it kinking. This happened, through my own incompetence, when we first tried to do this with much loss of humour on all sides.

We also have a discussion about the pros and cons of spraying a product called Frostec on the vines if there is a forecast of night frost. This contains a concentrated nutrient called Harpin whose application increases the natural defence mechanisms in the young buds from frost damage. Who knows whether it works but we do it anyway.

Of all the manual tasks in a vineyard, bud rubbing – removing the young shoots and buds on the trunk of the vine –must be one of most back breaking and sends a shiver down any vineyard labourer's spine. We have 1,500 vines to do in the Chardonnay field so Capability Louth puts his mind to devising a tool which looks much like a bog brush on a long stick with which one can rub up and down the vine inside the rabbit guard without taking the thing off. I have my clipboard and time watch as Tony and Robert creakily work their way down each row. I make a mental note not to volunteer.

June 2015

Blue skies and warm weather at night mean that we can remove all the bougies from the vineyard and store for next year. A lot of trips are needed to the store with barrows full of these very messy & half used tins. Gloves are essential. We should have enough to last for at least 3 years.

The vines are powering vigorously towards the sky and we need to keep on top of them. We spend 3 days moving from row to row carefully pruning each vine. I wonder how they can do this effectively in a commercial vineyard. The answer is with a large gang of labourers, yet it is hard to imagine that they can spend much time on each vine. Hopefully, we will speed up as we get more experienced.

Flowering. If you don't know what to look for the flowering process can happen right in front of your eyes without you realising. Depending on the temperatures, flowering usually happens 40-80 days after bud burst with small clusters of flowers appearing on the tips of the young shoots. A few weeks after the initial clusters appear, the flowers begin to grow with individual flowers becoming observable. It is during this stage of flowering that pollination and fertilization of the grapevine takes place with the resulting product being a grape berry containing 1-4 seeds. So, there you have it! Now vitis vinifera grape vines are hermaphroditic with both male stamens and female ovaries being able to self-pollinate.

Robert and I are out with the reference books and do site inspections every day in a bid to work out what is happening under our very noses!

16 June. Appointment with an officer from the Wine Standards, an off shoot of the Food Standards Agency. We are obliged to register the vineyard and have done this already, so now they just want to know that we exist, and more importantly when exactly we plan to have our first harvest. The young chap is called Jindrich Sedlecek and he covers all of Berkshire and surrounding counties. He seems impressed but does not have much knowledge of viticulture. The production of wine is highly regulated in this country and both the Wine Standards and HMRC take a keen interest in making sure all vineyards and wineries make accurate declarations.

20 June. Who is going to design our wine label? The most important part of the bottle other than the contents. We have gone slightly 'left field' and hit upon, via Gordon my brother, a fine typographer who resides in Edinburgh by the name of Robert Dalrymple. He does a lot of work for the Fine Art Society in Bond Street, and, in fact so it seems, most art galleries south &

north of the border. Rob is up for the challenge and sends me a wine merchants' catalogue he designed a wee while ago as a sign of positive intent. This is as close as he has got to designing a wine label.

I send him a specification of what we have in mind: I trawl all the UK vineyards' websites for elegant fizz labels and send him digital copies. I also include a sample of the also-rans, which make up the majority of UK wine labels alas, just to indicate what we do NOT like. For good measure, I include a few of the French labels, like Moet and Tattinger, distinguished by their red, green and gold colouring. The UK sparkling labels are distinctive yet understated, with sparse typography, which by and large I prefer.

Rob Darymple calls back. Gosh, there are a lot of rules and protocol to follow when it comes to designing a wine label, he chortles. And so there are:

- The front label normally displays the alcohol content (although some put this on the back), whether it contains sulphites, the quantity, the derivation and of course the wine name.

- The back label is customarily used to provide the background to the wine, the grape varietals, the name and address of the vineyard, the terroir (soil, topography and climate), and sometimes the maker of the wine.

- The neck label, which sits below the foil, is equally critical to the look and feel of the bottle; it usually has an oval or round rose at the front and back of the band on which the vineyard's logo is normally applied.

July 2015

Key task of pruning back the leaves on the Chardonnay vines as they are getting quite shaggy. It is important the bunches are exposed fully to the sunshine and not shaded by leaves. Takes longer than we imagined, well that's a first!

10 July. We planted the Pinot Noir vines in 2014 so this is their second season in the ground. Now is the critical time to select the primary shoot as we want all the vine growth concentrated on this one cane; if we don't do this now, we will have missed the best part of the growing season. It's all hands to the pump and we pull in Tony and his son, Martin, to help. Back-breaking work that is best done on one's bottom shuffling along the ground. At the same time, we slit each rabbit guard open with a knife, do our rubbing, and then put back in place with a cable tie around so in future the task of bud rubbing will be facilitated by simply untying the cable tie or sliding it up the stem.

20 July. John Buchan has impressed on us that we must keep the leaf vigour of the vines in check and that means topping each one just above the top wire – 'we are not growing a hedge', is his mantra – and making sure there are no stray shoots which can be decapitated by the tractor and sprayer as it comes down the row. Capability Louth does both fields methodically as always.

29 July. I source the correct length cable ties, second time round, from a website called Just Cables. These arrive the next day and I set about tying ties around 2,500 rabbit guards. It will be some years before we can safely take them off.

31 July. Major canopy task: de-clustering the grape bunches and de-leafing around each bunch to afford plenty of sunshine. We are told by John to select the best bunch on each shoot – ideally the one nearest to the laid down shoot and discard the rest. This is heart- breaking to lose all these bunches but you have to be cruel to be kind and avoid stressing the young vine by letting it over-crop. It takes 3 days to do 27 rows. Now that is dedication and I wonder how they do this in a commercial vineyard, or not!

August 2015

After a wonderful hot and dry July, we enter August with heavy rain showers and the long-range forecast is for unsettled weather throughout the month. August has had a record of poor weather for many years and I feel sorry for all those holidaying under canvas. Robert is back in Greece sweltering in the Cyclades on his boat Ria and wishing for gentle winds to cool things.

3 August. Borrow a hedge cutter from A4 Hire to give the vines a haircut. The trick is to walk down the vines slowly, holding the cutter blade at belt height horizontally to the vines, and cutting smoothly as one paces slowly down the row— not easy as the ground undulates. No one admits to cutting a wire but just the odd grazing here and there. Eventually CC goes for broke and severs a top wire – PING!

14 August. This rain is stimulating the grass and weeds around each vine. In between the showers I get out with a knapsack sprayer and apply Diquot. Each row takes about 15 minutes. It is a contact herbicide that seems to have little effect on the grass……or weeds for that matter. I might even mix it with my G&T it is so mild.

18 August. John Buchan drops by to cast his eye over the vineyard. He thinks we have a mineral deficiency in parts of the vineyard which is showing in some of the leaves – quite common and something that can be addressed through regular granular application in the winter with boron, magnesium etc.; in the summer one can apply foliar feeds. We agree to do a 'random' leaf sample and send off for analysis.

A lovely day, at last. Tony comes and sprays, as he has been doing every 10 days since June. This is not a job for the amateur as it requires maintaining a constant speed down each row, having the correct nozzles open as you start each row, and with full concentration that you don't miss a vine — not easy with the orange bucking bronco of a Kubota. Tony is a master tractor driver and has got the hang of her ladyship. She is strong-willed.

We have been following the spray programme guide issued by the UKVA (UK Vineyard Association) to all its members but John Buchan adapts it according to the conditions. He sends us small bottles each week made up in brown parcels. Given our bijoux size we only require tiny amounts of pesticides that only John can decant for us from his 'laboratory' at home. Next year we will get more organised, says Robert.

August has given us 3 weeks of God-awful weather such that the poor vines have hardly seen a ray of sunshine for more than 2 hours at a time each day. This bodes badly for the swelling of the grapes.

September 2015
The first week of September is much like August: raining cats and dogs with cold temperatures. I cannot believe that the night temp last night went down as low as 4.4°C.

7-12 September. Plumpton College, South Downs, Sussex. Centre of Excellence for Wine Education. Off to Plumpton for the week. Although we will not be making the wine ourselves we feel that a sharp intensive course will set us up in the knowledge department and of course help us influence how Emma Rice at Hattingley Valley makes our fizz – or at the very least comprehend her methodology. The weather is perfect with bright sunshine every day so our vines should be basking in the early September sunshine back in West Berks.

The students for the 'Principles of Wine Making' intensive course all gather at 9.30 sharp on Monday morning in the lecture theatre. Our lecturer for the week, one Tony Milanowski (an Aussie), begins by asking each in turn – a good ice breaker this – what was the last bottle of wine each of us drank. We have a round of dull New World Pinot Noirs and then Robert chirps up 'A bottle of Pavillon Rouge du Chateau Margaux 2001'. This was consumed in our hotel restaurant last night having brought it from Roberts's cellar in swaddling clothes. R told the hotel cheekily that as part of our wine course we were doing tastings and thus needed to do homework in the hotel; they took it hook line and sinker… and no corkage. Tony pauses and a balloon comes out of his head along the lines of, 'Strewth, that is a humdinger of a bottle, who the feck is this fella drinking Margaux as his vin de table.' Robert looks quite smug.

The students for the week are a mixed bag: retired doctor, an Australian barrister with a vineyard in Adelaide (we told him this did not count) who wanted to swap it for one in Italy, bio-dynamic enthusiasts currently lecturing as landscape designers, wine 'educators' on the London Eye, Lincolnshire farmer who has eyed an opportunity to increase her land value possibly by planting vines, an ex-mountaineer an IT specialist with a vineyard north of Inverness which made our enterprise seem rather tame, and a lovely lass working in the family wine supply business to off and on trade who has a plan to plant a seriously large vineyard in Dorset. And I nearly forgot Neil, the delightful chap who made chocolate machines for the Germans and who was only doing the course so that he could 'mug up' and enjoy time with his ex-pat son, whose passion was growing vines in his walled garden back in Leicestershire – that is when he was over from Brussels where he worked for the Commission as a lawyer! I think we confidently call ourselves a mixed bag.

Tony Milanoswki is quite a fella himself: a chemical engineer by training, he has worked in wineries in Australia and Italy with 12 vintages under his belt. He had the misfortune to end up in England having married a 'Pommie Sheila' but takes every occasion to moan about this wet and crowded island. Hampshire comes under most stick as a county with no roads going through it. He must be related to my postman in Hungerford as they both wear Dr Martens and shorts in all seasons.

Plumpton has become, in the space of 10 years or so, the centre for the study of viticulture and oenology in the UK; their undergraduates and post grads come from all over the world and then filter out into various wine fields. Emma Rice, our wine maker (and UK wine maker of the year 2014), is herself on the list of glittering alumni from Plumpton College. It also runs a semi-commercial winery making 50,000 bottles a year of sparkling and still wine from its own estate.

Tony teaches us each morning in the lecture theatre, with mini breaks every 50 minutes, and makes it fun with his little reflections on English life (Bill Bryson-like) and stories of making wine in extreme climates. He manages to impart a huge amount of information, much of it very scientific, to an audience many of whom — like us — have not been in the classroom for 40 years plus. Clearly, making wine is both a science and an art but, boy, I wish I had paid more attention in chemistry lessons at school. Robert is in his element but I find myself struggling with the scientific concepts and complex processes.

The afternoon sessions are spent in the winery and the adjacent laboratory with Sarah Midgley, the winemaker, who tries to teach us the wine making processes: from operating the de-stemmer, pressing the grape (rejects from Waitrose) in a bladder press, pumping the juice into a tank, analysing the grape juice, and doing various filtering to remove solids. There is a lot of childish sniggering and eye rolling with many of us standing around waiting for instructions that never come. But a pleasant way to spend an afternoon.

October 2015

The beginning of October brings continued sunshine and our fingers and toes are crossed that the grapes will speed up their ripening.

3 October. Our first vine leaseholders' lunch was held in the barn with a very good turn up of some 30 guests. I should here explain who these vine lease holders are. They comprise friends who rashly agreed to part with their money and take a punt on our being able to produce wine sometime in the future. In return they will receive large quantities of the first few vintages free of duty and at a discount if all goes well. There is nothing like organising a party and then failing to turn up oneself. I had been laid low all week with a nasty virus and by Saturday had still not managed to shake it off, so regretfully, on what was a glorious early October day, I stayed in bed and nursed a migraine. Robert was left to hold the fort and did an admirable job; he initiated proceedings with a lecture tour of the vineyard before lunch and answered, on the hoof, highly probing questions from several of the vine lessees. Bright lot these wine buffs, you can't pull any wool over their eyes!

Simon Taylor, owner of Stone Vine & Sun in Winchester, kindly agreed to come along and talk about the wine he supplied for the event. Naturally (and most diplomatically) he chose an excellent English fizz from Hugh Liddell of Cottonworth whose vineyard is close by to SVS in Hampshire. This fizz is made by......you guessed it.... Emma Rice at Hattingley Valley. Jeremy and Serena Hawkins very kindly contribute one of their lambs which Jeremy barbecues to perfection on a charcoal BBQ situated just outside the barn doors. Situation is key so he does not miss out on any of the jollity or suffer a momentary empty glass. Leonie and Frankie Louth effortlessly prepare the remainder of our sumptuous fare and I believe a lovely day is had by all (except for yours truly).

5 October. A visit from John Buchan to check on berry veraison progress. The majority of the vineyards he visits are at least 2 weeks behind in terms of ripening – some comfort – so he estimates we will not be picking until the end of the month. He inspects for disease but finds all is healthy. We do a Brix test for the sugar levels and record a brix of 11.4. The target Emma is looking for is close to 16! The legal minimum for natural alcohol in wine grapes as set by the Wine Standards is 8% so we are not there yet.

8 October. Robert takes a random sample of grape berries, puts them in a plastic bag and takes them down to Custom Crush at Hattingley Valley. They do the analysis there on the spot which is mighty efficient, and we leave with a printout of the results … plus a nod to come back with better results. R was told to bring 300 berries but, through a misunderstanding, only takes 30 so gets a bit of a gentle ticking off! The key measurements are as follows: The Brix is 11.1, the T/A acid level (titratable acid) is 25.5, and the pH is at 2.83. The lack of sunshine during August and early September has badly affected the acid levels in the grapes so we are well behind where we should be at this time of year, i.e., 2 weeks away from harvesting. Ideally the T/A level for harvesting fizz grapes needs to be in the region of 16 – 18! Therefore, still some way to go.

14 October. On this occasion we both take separate samples of berries and then test the sugar levels on our new refractometer (expensive but v useful). We get different results – mine is higher by 1 Brix point!

In the meantime, we are noticing increasing signs of botrytis in the grapes and Robert embarks on the tedious but essential task of removing as much as we can from the affected bunches; whilst doing it with utmost care so as not to do any further damage to the bunches. It is also clear that something is nibbling the low hanging bunches. It is the local pheasant population that have started to trot into the vineyard to supplement their diet of corn. Therefore, it is decided to install a gas gun at the top of the vineyard to scare them off. To increase the decibels R places an empty drum a few yards away from the end of the gun. When it goes off it sounds like the starting gun at Cowes Week and we both run for cover. They must be cursing us in the neighbourhood.

20 October. I drop in on Jane Thorp at Cobbs Farm shop

nearby where they have a 5-acre vineyard. This will be their third year of harvesting so she is an experienced hand. This autumn will see the launch of their first vintage with their Alder Ridge sparkling. In spite of the fact that Cobbs grows every kind of fruit imaginable Jane and her colleague Alison find grape vines the trickiest of all her fruity children. She reports that whilst the sugar levels of her Chardonnay and Pinot Noir are acceptable, the acid levels are still way too high. We have a small joint moan.

24 October. It is my turn to take a sample of grapes down to Hattingley Valley. I arrive early and Jo Tommony is already in the Custom Crush laboratory, a converted farm building which sits in the corner of the yard. She takes the grapes and crushes them inside a plastic bag until there is a goodly amount of juice created which she extracts with a syringe and then goes to work on the lab bench. I wait with bated breath. The Brix is up to 14.3 (identical to reading from the refractometer at home), good news, but the tartaric acid is still too high at 19.5 but coming down. She gives me that sympathetic hospital nurse look. Let's wait another week.

30 October. Red Letter Day – The vendange. The season has become decidedly autumnal, but we have a small weather window. Having consulted with Emma at HV, where they have been taking in grapes now for over a week, she looks at her timetable of slots on the wine press and the die is cast. We must go for this Friday if we are to have a chance to harvest.

Robert and I spring into action. Calls are hastily made to local vine leaseholders and friends with a horticultural bent. Within 20 minutes we have gathered a gang of pickers.

My partner in crime 'prepares' the barn like a surgery: plastic sheets on the floor, various benches in line, a quantity of buckets, and vine secateurs by the dozen laid out for the pickers. By 9 am the pickers have all arrived and are assembled in the barn – dressed in wet weather gear with buckets and secateurs at the ready – for the pre-harvest team talk given by the two vignerons. Each picker is allocated a row and instructed to cut out where feasible any berries on bunches which have signs of rot (botrytis) as they go down the row. Bunches are placed in buckets which when full are brought by Leonie, Frankie and Theo up to the sorting tables in the barn, manned by Robert and me. For the first hour things are slow with a multitude of questions from pickers but after that a good rhythm is established with less chat, and the buckets begin to come in steadily. Each bucket is tipped carefully on the tabletop, and we sort through, rejecting any bullet bunches (very few) and snipping out the odd rotting berry, which has been overlooked.

There is no uniformity in the quantity of grapes picked from each row; the young vines are at different stages of maturity and some canes were not thick enough to lay down on the fruiting wire, albeit they still produced a few bunches on the vertical, so it is hard to estimate the length of time it will take to pick. For instance, row 1 has 173 bunches whilst row 23 is a monster with 367, and the latter takes 4 pickers over one hour to harvest.

One of our pickers is more accustomed to harvesting in the Languedoc where you are down to shirtsleeves by 9am and, for protection from the beating sun, a hat is de rigueur. I fear cool climate harvesting is rather a different experience. At least the predicted rain has stayed away.

After the first couple of rows are harvested, we have hardly any grapes and Robert and I look at each other with a certain apprehension: help, are we going to fill the trailer at this rate. But after the second hour the buckets pile in and the crates begin to multiply nicely. Robert moves them to the awaiting trailer that is parked outside the barn doors and neatly stacks them.

Lunch is a hurried affair as we are up against the clock having underestimated how long the process is going to take. Some folks, who we only booked for the morning, have pressing commitments so down tools at midday — luckily to be replaced by others. A couple of hurried calls later and Doug Garrod and Nick Boden appear to lend a hand.

A big round of applause for our harvest helpers who all did sterling work: Doug and Christine Garrod, Tina Lowe, Kate Rossiter, Chris and Bridget Roupell, John Burleton, Jeremy Hawkins, Sirio Quintavalle, Tatiana and Becky Quintavalle, Gillie

Smith, Leonie Cooke, Frankie Louth and Nick Boden on drums.

By three o'clock we are done… and bushed. The last crate is stacked on the trailer and Commodore Louth lashes them together in fine nautical style so that even my erratic driving on the A34 will not affect their stability. We celebrate with cakes, coffee, and beer – no time for a leisurely lunch, as we must get to the winery for our booked slot on the press.

We arrive at Hattingley Valley by 4.30pm. We are the first vineyard to deliver that day. A welcoming party materialises from various buildings as we draw up in the yard and lots of jolly young men start to help get the crates off and onto a waiting pallet. They inspect the crates of grapes and make appreciative noises. Emma Rice appears in her fetching cowboy hat and mans the electronic scales.

Within half an hour we are back on the road with an empty trailer heading home, both feeling slightly lightheaded after the day's exertion. After 3 years of back-breaking work in all weathers, our first crop of fine Chardonnay berries is finally harvested. We are now completely in the hands of our winemaker.

2015. Robert and Christopher in the lab at Plumpton College

2015 April. Bougies burning bright

2015 April. The morning after the frost

2015. Robert on the orange beast

2015 August. Both fields early morning

2015 October. Trailer full of Chardonnay

2015 October. Christopher on sorting duties

2015. John and Robert digging for soil samples

Christopher harvesting

2015 Harvest - Tina

2015 Harvest - Chris

2015 Harvest - John

2016 March. Jeremy and Chrissie

2016 Harvest. Robert, Leonie, Louise and Richard, crates at the ready

2016 Harvest. Tania and Caroline 'buddies'

2016 October. Harvest. Bridget with her snippers

2017 Harvest. Penny

2017. Harvest. Leonie snipping

2018 Harvest. Sonja and Louise

Our Second Year of Production

February 2016

As we enter what will be our 4th year in the vineyard and our 2nd year of grape production, it is frustrating to think that we are still 24 months away from the release of our 2015 vintage. Our bottles sit on their lees slowly maturing in a darkened room at Hattingley Valley winery. You need to have patience of a saint to play this game. In June we will have a chance to have our first tasting from the bottle. No garlic the night before as we will need a clean palate.

It is a chilly February day as we gather in the vineyard with Tom Bartlett and his replacement called Jake, both of VineCare, for a practical lesson in how best to prune a vine in a single guyot system (i.e., one cane). Jake is taking over from Tom who, as assistant wine maker at Stopham Vineyard, is juggling too many duties to continue touring the country as a peripatetic viticulturist consultant. He has been great. Firstly, we have to cut away all the dead wood from last year's growth which involves disentangling all the incredibly strong tentacles which the vine throws off to grip the wires; next select a new cane with 12 potential buds plus a spur from the trunk which ideally is located 6 inches below the fruiting wire so as to make bending onto the fruiting wire simple. The spur will become next year's cane. Their advice is to wait for the cane's sap to drain away before we actually tie the cane down so as to avoid any disease seeping back down into the trunk. This will be a job for early March.

It is that time of year to get out the thermostats and place the bougies on guard like sentries at the ready. Gosh, it really is a filthy job and we get covered in dirty bougie wax.

April 2016

A critical month when night frosts can kill the tender vine buds and ruin the whole crop in minutes. On 13 April we spot the Chardonnay buds beginning to change their characteristics and move from dormant to woolly white. On the 15th we spot the early stages of this also in the Pinot vines.

16 & 17 April. We experience heavy frosts, as predicted by Metcheck, and we are up all night lighting bougies in the Chardonnay field. We had installed heated wires in the Pinot field but not the Chardonnay. It falls to 2°C by 8pm and the heated wires in the Pinot wires, which are set to come on at 2°C, perform perfectly and are on by 8.30pm! There is reported devastation around the country and in Europe too. Bush telegraph moves fast. We get calls from fellow vignerons as to how the heated wires are functioning. And 'Do you have any bougies for sale?'

Temperatures can drop in a matter of minutes to dangerous levels. Dawn is known to be the coldest time – just before the sun rises. Air frost is more critical than ground frost largely as the fruiting wire is 80cm from the ground so mowing the grass low in April is advised to reduce the ground frost on the grass.

There are two main types of air frost: radiation and advection. Radiation frost is rapid heat loss from the earth's surface that occurs on calm still nights. Advection is 'the result of a large cold mass of air moving into an area' – definitely the hardest to manage and control.

In 2016 many vineyards in France suffered badly from spring frosts and Champagne particularly. Vineyards in Austria too were devastated. So it is not just England that gets badly hit. Parts of Hampshire experienced -5°C degrees on one night in April.

May 2016

The call goes out for a bud rubbing gang and on the appointed day we have a gathering of 'the faithful'. Slowly we work our way down the rows, on our knees, removing each vine guard and rubbing away any buds that tend to sprout on the trunk of the vine. If we don't take them off now these will become shoots and cause us a nuisance later – we only want growth through the cane shoots.

Debbie Higgins has dressed in a sou'wester as if she is either volunteering for the Salcombe lifeboat or going to Glastonbury. But she has the last laugh as the heavens open and we get drenched.

We have a bad frost forecast for 13 May and we keep an eye on Metcheck. The trouble with frost is that it can be very localised and we know for a fact that our site is always colder than Kintbury down the road. The bougies in the Chardonnay field are in position.

By mid-month, as night-time temperatures begin to rise, we decide that the time is right to attack the final prune. On the advice of Tom Bartlett, we left our canes extra-long to protect against frost. Has it worked? We don't think so. As a result, we have another procedure to undertake which will take the best part of 2 days as we have to go to each vine and remove excess buds: 2 or 3 buds depending on vigour of vines. Spur selection is critical, as this will provide the cane we lay down next year. We have left 2-3 buds as a precaution against frost but we will select the chosen shoot later on.

We suffered some vine casualties last year and these we will replace with new vines. I work my way through making a count, mark with a clothes peg, and then put an order in for 20 replacements.

By the end of the month it is possible to count the cost of the damage by seeing how many buds have made it and those that have been zapped by the frost. In the Chardonnay field it looks as if we have lost 25-30% thanks to that bad night when even the bougies could not hold back the heavy frost. As for the Pinot Noir vines which were less advanced when the heavy frost hit (at woolly bud stage) but which also had the protection of the heated wires, we have lost almost nothing. Food for thought.

June 2016

Weather-wise, the first week of June is fabulously hot. Long may it continue. By the 12 June we collect up all the bougies from the vineyard and store. The empty ones are taken on a trailer to the metal recycling depot near Bradfield which is full of old cars being crushed. I park the Land Rover carefully as far from the crushing machine as possible. We receive a miserly £2.70 for our 200 empty tins. I will not bother again as I must have used £20 of fuel to get there.

20 June. Robert returns from a month away sailing in Greece. I pass the baton and he starts working through each row taking out blind shoots (no bunches) and tucking in rampant shoots so they are not swiped by the sprayer as it runs down each row.

July 2016
This is the month when major canopy management is most required. Sunshine and rain have made everything grow madly.

12 July. There is lots of tying back of rampant shoots in the Pinot field. Next job: take off 2nd bunches on secondary shoots as they will not catch up. We want all the ripening to happen at the same time.

13 July. First hedge cut of the season. Boy it is hot. Wimbledon finishes and finally the skies clear with fab weather. Record 33°C today and it barely drops below 28°C for the entire week.

21 July. Another visit from John Buchan. Bunches are setting well which means we should not have the same problem we had last year of uneven ripening (I will hold him to that!). He reports that he can reduce acidity levels with the addition of nutrients. We will hold him to that too! High acidity levels in England are the big challenge to a wine maker and late harvests are common as the grapes need long periods of warm temperatures. Last year we harvested right at the end of October.

August 2016
We have something called rain today. All very odd. Nevertheless, temperatures still hover around 20°C. Chardonnay: we start thinning the leaves, topping and taking out laterals yet leave 2 leaves so it can re-grow. Hot weather continues unabated, hurrah.

September 2016
Pinot bunches are really turning red – and startlingly so. Hot weather continues and hovers around 25°C all week. Lovely for the vines. Lots of green harvesting to do in each field and we spend all week doing so. Hard work. This is watch and wait time while we keep the canopy in good order and the rides mown.

A short sojourn to the south of France to celebrate our thirtieth wedding anniversary in the Jaguar soft top. It purrs all the way and does not miss a beat. First stop is Épernay in the heart of Champagne, then we wind our way down to Beaune in Burgundy; and then finally stop just south of Lyon for the night. We have arranged a tour of a few Northern Rhone vineyards in St Joseph and Côte Rotie with an

Englishman who was a tax accountant in a former life, then became a wine supplier; and now has gone most definitely native. Without his contacts we would never have had this sort of access – wonderful.

Down to Hattingley Valley with grapes samples. We will use this sample as a benchmark: the Pinot is miles ahead of the Chardonnay in terms of its acidity levels as measured by T/A (titratable acidity) as the Chardonnay has a whopping 27 T/A pts! A very long way to go.

October 2016

This is the build-up to harvest thus lots of samples and analysis to make. Our methodology when sampling is to ignore the outside rows and the beginning of each row as they tend to have the least ripe bunches. Walk down the row, close eyes and pick off a berry and pop it in the plastic bag. 300 is a good sample number. We use a Brix test with a refractometer to test the levels of sugar but we don't have the equipment to undertake a T/A test as this requires laboratory equipment.

Vendange day for the Pinot is 12th of October as sugars and acids are in perfect harmony. Hattingley Valley winemakers say Go! Last year we borrowed crates from a local nursery but this year we have been on a buying spree and located crates for sale in Devon from chap who sadly is selling up his house and vineyard after 12 vintages. Astonishingly he grubbed up his vines in order to sell the house. Additionally, we borrow another 100 green nesting crates in Henley from a chap growing still wine in a vineyard planted by his mother 30 years ago. The vineyard looks a bit unloved as he lives in London and comes irregularly to spray and tend.

The calling cry goes out and we gather a posse of pickers. A dry day, thank goodness. We start with a Robert lecture on what to pick and what to leave aside. Then, off we go. By 1.30pm we are all done with 600 kilos of beautiful black grapes glistening in their crates.

Now in the 3rd week of October and the ripening of the Chardonnay bunches is proving difficult. It is a fight against disease setting in whilst wanting to leave the grapes as long as possible to lose some of their high acidity levels. Emma wants a T/A of 14-15. To that end we do a botrytis patrol each day and take off anything infected. To make sure we prevent any spread we cut out any berries close to any rotten ones. On average we take off 2 kilos per circuit. In between we spray with a product called Serenade which works by eating the bacteria within the bunches; we also use Foli K which contains potassium for reducing acid levels.

The weather has been like an Indian summer for most of October with temps of 14-16°C most days plus warm nights which are key as ambient temperature is what the grapes need.

21 October. Another beautiful day in the vineyard with max of 19°C

29 October. We strip at least 15 kilos of bunches either infected with botrytis or bunches with shrunken berries and fungus on the stalks. Very bitter on the tongue. Largest amount to date.

31 October. Very hot at 18°C. Our Brix is 17.2 which is fine. Another 15 kilos taken off.

November 2016

We take down another sample of grapes to Hattingley Valley: their advice is to pick now as frost is predicted this week which could wreck our harvest if we don't pick. We get on the dog and bone forthwith and assemble a team for the morning. Test results: Brix 17.1 and T/A 15.9. The goal has always been 15 T/A or under. The sample of 300 berries is taken from both sides of the vine and it is clear that the east facing bunches are less ripe than those facing west.

2 November. The Chardonnay harvest starts at 9am sharp with 12 pickers. It all goes to clockwork and we are finished by 1.30 pm with 84 crates on-board the trailer – slightly less than we had estimated. Leonie and I are booked on a flight out of Heathrow for New York at 4pm so alas we miss out on the harvest lunch which goes on through the afternoon. It is a good feeling to be done and dusted before the frost. It is over to the winemakers now. They analyse the grapes and the final Brix comes out at 16.5, T/A up slightly at 17.3 and pH 3.0. Our tonnage from the Chardonnay is 1,248 kilos.

2016 March. Christopher and Robert - double act of pruning

2016 June. Debs and Chrissie bud rubbing

2016 June. Young Pinot field

2016 July. Partners do canopy work

2016 August, Christopher snipping

2016 August. View up a row of Pinot

2016 October. Chardonnay bunches

2016 Octobe. First batch of Pinot

2016. Robert prostrate with Tattie for company

A Season Dictated by Spring Frost

January 2017

Another year, and another growing season to think about. We have not touched the vines since harvest last November but winter pruning cannot be put off for much longer, particularly as we are blessed with a period of ten gin clear blue days in January when it is a pleasure to get out in the fresh air and get those secateurs clicking. The old cane is cut off together with last year's shoots leaving just this year's cane standing vertical from the trunk. It is advised not to lay the cane horizontally at this time until much of the sap has drained out in case of disease transfer to the trunk. November's light herbicide spray of the grass amongst the vines has proved effective and the rows look clean and tidy for the most part. We create an enormous pile of vine cuttings in the field beside the vineyard which we will torch when tinder dry.

On the agenda of items to think about: a new air-assisted sprayer which we believe will give better coverage notably in the latter part of the season when botrytis can set in. This will then mean a new tractor, as the present one only has 12hp – not sufficient to drive a sprayer of this type. Robert wants to sub soil alternate rows to break up the soil around the roots of the vines that will have become, in all probability, compacted by running the tractor down the rows over four seasons. That presents a technical problem as most sub-soilers are beasts of burden requiring at least 50 horsepower and we are limited by the 1.8m width of our rows. He likes a challenge.

February 2017

Robert and I sit down and discuss the feasibility, timing and cost of installing heated wires in our Chardonnay field.

The quote from SSE to bury a power cable 70 metres comes in at £3,000 so that decides the issue: we will bring armoured cable, some 500m, from each heated wire circuit (210m each) back up in a trench to the meter box. We go for the cheapest option from SSE which is to drop the cable down the pole and into a box at a cost of £300.

The search for another compact tractor begins. There are lots to be bought from Germany but how the devil do you know what you are buying. Then there is the transport cost. Eventually we get a call from John Buchan to say that there is a Kubota with cab for sale from an organic vineyard near Padstow. I call the owner and ask him various questions. Clearly, he is not technical and Robert takes over and speaks to his mechanic. They want £8,000 for something that has done 4,000 hours. It sounds perfect but then there is the headache of getting it back to West Berks. My trailer is too short so we borrow a ginormous one from Kintbury Holt farm (this was used for the Chardonnay harvest) and hook it up to R's BMW.

We arrive at Trevibban Vineyard, armed with £2,000 in cash and a computer to make an instant bank transfer for the remainder if we like the machine. Engin is Turkish with an un-pronounceable surname. The Ottoman and the British empires lock horns in one almighty haggle over £250. Eventually a shake of hands followed by a 2-hour lunch with Engin and his delightful wife in the vineyard's fantastic first floor restaurant – called Appleton's. We make it to Taunton that night and crash in a heap of exhaustion (and alcohol). The following morning in driving rain we negotiate our precious cargo back up the M5 and M4 – avoiding jarring potholes in the motorway – and it is with some relief that we reach Orpenham by lunchtime, tractor intact.

The case of the thieving blackbirds. Last summer when the Pinot berries were sweet and luscious, we had blackbird raids from the hedges on either side of the field. These hedges were predominantly hawthorn thus popular with blackbirds as it provided a safe haven from predators. They stripped the outer rows at will and started to make inroads on the next ones in before we cottoned on to what was going on. We think we lost 15% of the crop just to bird damage. The only solution we decided was to take both hedges out, requiring a 3-tonne digger and dumper. The result was rather good and such was the width of this old and scrawny hedge that we have created space for another 3 rows of vines without impinging too far onto the drive lawn. Another 150 vines planted will in due course provide a further 250 bottles! Worth the investment.

March 2017

As expected, this proves to be a very busy month in the vineyard: we have the installation of the heated wire system in the Chardonnay to manage: 1.5 kilometres of heated wire to thread along the fruiting wire, not to mention 200 metres of trench to dig in which to lay the armoured cable to the electric box. We hire a trenching machine and Robert can't wait to work out how to use it. Fortunately, he has plenty of room to practise as it becomes apparent quite quickly that you pull it backwards not forwards! Stones and soil spew out as he works methodically steering the beast down the field.

It takes us 3 days to install the system and 3 days to clear up the trench mess and pick up the flints. The electricians, having done one field last year, move fast and encounter no problems. Thank goodness we will not have to do this again.

The next important task is to lay down the canes onto the fruiting wire. As per last year the Pinot canes prove a doddle to bend and tie— we are done in 2 days. Where we have short inter-nodal distances and too many buds, we put on clothes peg to remind us to go back after the frost danger period is past and reduce the bud number. This year we will lay 9 buds down for most but just 6 where the vine is being laid down for the first time.

As expected, the Chardonnay vines prove to be brittle and non-compliant to our entreaties to conform. Tony gives us a master class in how to twist and bend without snapping but even he manages to snap a few. When the canes are wet with dew or rain they bend beautifully. We toil up and down the rows for several days — 250-300 vines are about the max one can do in a day if one is being ultra-careful. I develop a system of carrying a knapsack of water on my back and spraying a section of cane vines at a time; then re-trace my steps and begin the bending process. Just enough water and they become pliable. On the final day I have my Eureka moment: if I take off all the rubber ties that attach the trunk to the metal vine tutor – this keeps the vine growing upright – I can then manoeuvre the cane to the horizontal with a smooth bend rather than an acute one. It seems to be the trick but of course it takes time to reattach the rubber ties afterwards. Only snap 2 before the end by being overconfident and not listening for the cane groaning under the strain. Each cane has 3 ties around the combined fruiting wire and the heated wire and one can slowly squeeze the cane so that at least 90% is in direct contact with the heated wire.

Again, we use clothes pegs to remind ourselves of the need to come back to some vines and carry out bud reduction after the frost. Robert manages to run Hungerford dry of plastic clothes pegs on one buying splurge!

April 2017

The trouble with unseasonably warm weather in April is that it prompts Mother Nature to wake up prematurely and promote early vine growth: early budding can be dangerous as we know only too well, as any subsequent frosts can wreak havoc on tender buds when they are at their most vulnerable. On 9 April we registered 20°C at midday in the vineyard. Very warm.

We estimate that we are maybe 10 days ahead of last year in terms of bud development. On 3 April I awoke at 6am to see a spotlight on each field: this signifies that the temperature is +2°C or below as the heated wires have clicked on. Later I check one of the outside thermostats hanging in the vineyard and indeed the temperature fell to -1°C degrees overnight.

Paul Woodrow-Hill at Vinecare tells us this is the earliest bud burst he has witnessed in 30 years of working in the UK viticulture business.

Finally, after many months, we have a new working granary with running water (and a fridge!), which will become the vineyard equipment store, workshop, and canteen. Titch and his gang of carpenters have done a grand job in re-building the granary utilising where possible reclaimed beams from the original. The frame is made of green oak, the cladding of soft wood, and this sits on 9 re-bedded staddle stones. The plumber has finally connected running water to the sink so now we have all mod cons and Tony can have a cuppa tea and heat up his lunch in the microwave.

The second part of our rationalising the storage of vineyard equipment involves building a new tractor shed to house our new Kubota baby and all the other bits of tractor equipment. In anticipation of Annabelle Louth's wedding party in the barn we move the two tractors out of the carport on 9 April.

2 April. As we leave to go on holiday we stop at the end of the drive. Bud burst. Frostec at the ready to protect the spurs especially against air frost. They are not in contact with the heated wires.

24-27 April. The extract below is from an article I wrote for the Thames and Chiltern Vineyards Association, entitled *Armageddon in April.*

In February we took the decision to sell off all our stock of bougies! High-risk strategy? Hopefully, not. At Winding Wood Vineyard we have decided upon another frost protection system: heated wires. Having seen an experimental installation at Ridgeview in Sussex and read about heated wires being used in New Zealand's cool climate vineyards to good effect, we decided last year to install the system along the fruiting wires of our Pinot Noir. However last year, as our Pinot buds were only at woolly bud stage when the worst night of frost hit they were not at their most vulnerable so it was difficult to be empirically accurate about their effectiveness. In any case, we lost no buds that year which was encouraging.

In March 2017 we took the plunge and installed the system in our Chardonnay field – this involved attaching 1,500 metres of cabling to the fruiting wire. Last year, with bougies as our only frost protection, the Chardonnay buds were pulverized. Of course, we are no longer spring chickens and resent getting up in the night to light bougies; and secondly, we are not convinced bougies work down to -5° C in advective frost conditions – even when they are set out at double the recommended rate per acre.

The heated wires are linked to a sensor in the vineyard, which in turn is connected to a thermostat in the electric box set to turn on the circuits at +2°C (a safety margin) and heat up the wires to 20°C. The wires and the cane are tied together so the heat can travel down the cane. Thermal images taken at Ridgeview show the cane surrounded by a halo of orange. We think we are the only growers in the UK who have fully installed heated wires in our vineyard not surprisingly we are beginning to get a procession of fellow owners wanting to visit and inspect.

How did our heated wires manage with the three Armageddon

nights in April? Well, if you had asked us on the morning of 26t April I would have said confidently that they performed extremely well over two consecutive nights of – 4.5°C conditions with minimal damage – maybe just 2% of bud losses. However, the 3rd night of 26/27 April was a perfect storm: we had hail and rain in the early evening with northerly winds followed by –4.5°C temperatures overnight and the conditions decimated our advanced buds. Clearly, the heated wires just could not cope with the freak combination of wet and freezing conditions. We estimate to have lost 70% of our primary buds which by the afternoon of 27th resembled dried tobacco plants. Given the unusual conditions in April this year everyone's vines were ahead of normal. We had an average of three leaves out – and I wonder if our buds had been less advanced we would have been so badly affected. Without wanting to indulge in schadenfreude, it was interesting that other vineyards in this area, which employed no frost protection save spraying Frostec, were 100% wiped out. Whether the secondary bunches will be worth having only time will tell.

It is too early to assess the damage but we suspect that the loss of primary buds amounts to 70%. Distressingly, one of the circuits in the Chardonnay fields — our best 3 rows in terms of bunch production — failed to come on the 3rd night for some gremlin reason and we have lost every single bud. We get the electricians to come in and check the connection, but they too are scratching their heads as to why this circuit did not fire up. It must be human error but one thing is crystal clear: the heated wires work.

More planting. We took out a hedge last year next to row 1 of the Pinot as the blackbirds used this as a launch pad for stripping forays of the adjacent row of berries. Firstly, we get Matt from Kintbury Holt farm to come along with his large tractor and break up the soil with a subsoiler and then Robert rents a machine to rotovate the 3 new rows. We then install the end and intermediate posts using a post basher slung on the end of our new tractor hydraulics with Robert cleverly turning the tractor on a sixpence with enough skill to avoid dropping the hydraulics down on my hand as I guide the thing into position. Boy, did it hurt, but no amputation required.

May 2017

Darling Buds of May? It's a waiting game to see what recovers from the devastating frosts of late April that pulverised our young buds. We are most interested in those primary shoots, which survived, as they will provide the best cropping bunches. The shoot will send out secondary and sometimes tertiary buds to replace any dead buds and of these the secondaries can sometime produce decent bunches of grapes. However, they may never ripen enough in our cool climates to be of use. What is clear is that we have inconsistent recovery: some vines have only two shoots whilst others have five or more.

June 2017

I have a gang of 4 volunteers (WWVVs) – Sue and Robert Bembridge and Jane and Martin Buckland – come in to help me. This is the first working party to go through the Chardonnay bud rubbing, loosening ties and tucking in any floating shoots. It takes all day for 5 of us to complete the job but for the moment there is nothing extra to be done until the vines have sprouted more leaf. On the following Sunday Heather, another volunteer, comes over for a morning session to finish off the Pinot. She soon gets the hang of things and by lunchtime we are done. All tucked in — for the moment!

As an experiment we decide to turn on a couple of the heated wire circuits to see if it improves flowering. We are told that vines shut down below10°C and stop growing; what is important for vines is the ambient temperature over 24 hours so it is no good having very hot days followed by cold nights. During early June I keep records of the lowest overnight temps and they range from 5 to 8° C. I gradually turn on 3 circuits in each field. We will see.

For some reason, lost in the mists of time, I stopped writing my 2017 diary at the end of July. However, I can report that the season progressed in line with the earlier ones. We harvested quite a small volume in October, thanks to the devasting frosts in April, though the winemaker was happy with the clean grapes we gave her.

2017 April. Robert with our new tractor bought in Cornwall

2017 May

2017 June. Full canopy

2017 July . Chardonnay in summer serried rows

2017 July. Great flowering

2017 August. Partners plus dogs walk the vines

Chapter Five

Frosts, Pests and Diseases

Despite our successes, England and Wales are still very much marginal areas for growing grapes. Southern England is at 50 degrees north latitude. Compare this to other wine areas: Otago which sits at the bottom of South Island, New Zealand is 45 degrees south latitude, the region for grape growing in Patagonia, is some 42-46 degrees south, although the country itself stretches to 51 degrees.

The biggest fears early in the season are spring frosts. These can, over a single night of minus temperatures, destroy young vine buds and wipe out much of one's potential crop. We have experienced Armageddon at Winding Wood over too many seasons not to have sleepless nights in late April and early May. Our vineyard is quite high at 130m above sea level and distinctly cold. We did not have much choice on our site but the location for planting a new vineyard is crucial: ideally a good south or west facing slope, situated not too high above sea level, and close to the coast – none of which apply to us.

Frosts

Bougies

One of the oldest methods to displace cold air is the use of bougies (candles) placed along the vine rows at intervals. These need to be lit at night by blow torch – a messy business – and at some speed to be effective. Bougies can last up to 8 hours of burning. This was our first line of defence in the early years, but it was very tiring to be on frost alert on clear night skies during April, especially as the coldest time was at dawn. We rigged up a thermostat alarm in the vineyard connected to our mobile phones which rang when the temperature tumbled to zero. Like firemen, we were dressed and out within minutes. The first year we performed this it was a thrill, aided by a tot of whisky. After that it palled. Not only are bougies dirty but they are environmentally unfriendly.

Air Turbines

An expensive method is to erect a portable air turbine in the middle of the vineyard, run by a petrol-fired generator. They are very noisy and unwelcome if one has close-by neighbours. These are used heavily in New Zealand, South Island (Otago mainly) where they suffer badly from spring frosts. Helicopters are another expensive option. They can hover over a vineyard and displace the cold air with their blades. A neighbouring vineyard, with endless cash at their disposal, brought in turbines during April/May of the 2020 Lockdown to run at dawn. It is a deafening noise. Nor surprisingly there was little sympathy from neighbours who had their windows open.

Sprinkler Systems

In recent years, there has been an increase in the use of low-flow sprinkler systems as a method of mitigating frost. They are not only very costly to install but are temperamental if not fully supervised. Water is sprinkled from overhead pipes over the vine which then freezes, creating a thin layer of ice around the tender bud. This produces latent heat around the bud and protects it. The system uses huge amounts of water – upwards of 10,000 litres per hour – requiring access to a built-in reservoir. The icicles around the bud are then slowly thawed once the threat is over.

Heated Wires

This was the system we decided to adopt after lots of research. Given that we are boutique in size, it meant that the costs, while high, were not off the scale. Considering that the retail value in wine of even our small plot is £150,000, it makes sense to spend money to protect the crop. While upfront costs are high, annual running costs are modest – after paying for the heated wires. A single-phase power supply is sufficient for the system to operate – akin to an electric shower. Once SSE had taken power nearby to an electric box on the edge of each field, we hired a trenching machine – a bucking mule which spews out earth and stones equally – to take the armoured cable to the end posts of the vine rows. 200 metres of trenches and 3 kilometres of heated wire. From there we divided the rows into several circuits, leaving the electrician to do his business in connecting the whole thing. Not for the faint hearted.

The wires work on a simple principle of conducting heat, similar to underfloor heating. In fact, the manufacturer, Gaia Climate Solutions, specialise in just that. A continuous heated wire is threaded along the fruiting wire and tied close onto the vine cane. It heats up to just 20°C, using no more power than an electric shower, and creates a halo of heat around the young buds to protect them from the swirling, frosty air. If you put your hand on the wire it feels warm to the touch. Any hotter and the bud would be zapped.

The Hungarian Frost Buster

At a regional vineyard event in 2023 at Whitehill near Lacock, the attendees were treated to watching a new frost buster, manufactured in Hungary, in action. The owner of the vineyard was a farmer in a previous life. He demonstrated the monster which is pulled by a tractor. As the tractor trundles up and down the rows the vast steam tank expels hot air over covering as much as 8 rows at a pass. The exercise needs to be kept up all through the night to be effective – yet a blight to the neighbours. The jury is out but it looks the business – with a princely price tag in excess of £20,000.

Pests

Deer

Which pest shall we start with? At the top of the chain must be the omni-present deer. At the beginning of the season, this is the biggest pest. Deer are smart grazers, checking out the best sites for feeding and then returning at dawn or dusk to take their pick of the young vine shoots. Muntjac are a nuisance but too short in the leg to reach our vines. Roe deer are enemy number one. I hope we never see a herd of fallow descend on us.

Many vineyards have deer fenced their entire plantations at great cost. It would be impractical for us, not to mention unsightly being so close to the house. Over the years we have doused rags in diesel to hang close by on spikes, used unwashed human hair (when we can get it from the hairdresser), until last year when we wrapped bird netting around the perimeter in each field as a deterrent. It seemed to work, if a pain to remove for early spraying. The camera showed little activity from deer.

Hares and Rabbits

There are plenty of hares running across the vineyard which is a pleasure. They are lousy parents, have no burrows unlike rabbits, and often leave their offspring to fend for themselves. Ludo, the vineyard spaniel, sadly has a field day in bringing me back days-old leverets. He has a soft mouth such that many survive and are placed back in the hedgerow.

The main casualty has been the roses Robert planted at the ends of the rows for show (not early sign of mildew which is an old wives' tale) – fittingly called Comte de Champagne: during the winter they were utterly destroyed by marauding hares seeking any form of nourishment. Grass is clearly not to their taste.

Birds

We have been reasonably fortunate in our experience of bird damage. When the grape bunches are ripe the bird kingdom are the first to twig, driven by the sweetness of the berry. The big threat would be a flock of starlings — although I have never seen such a sight in England – that are capable of stripping a vineyard in a matter of half an hour. Commonly vineyards hang bird netting along the fruiting zone of each row, either by machine or by hand. It is tedious but does the job. In the early days we found that cheeky blackbirds would hop out of a nearby hedge and feast on plump berries. Eventually we decided to take out the old hedge and plant several more rows of Pinot Noir. That did the trick with more bottles as a result.

Pheasants can be a pain, depending on the height of the fruiting wire, and are prone to wander into into the vineyard before the shooting season and take the low hanging fruit at will. By and large they are decorative and do little damage. Partridges are harmless. Pigeons are a pain by hanging and defecating on the overhead wires carrying the heated wires across rows – a potential problem for an oncoming tractor if the wire sags too low – yet there are never enough to do serious damage to the fruit.

For a few years we erected a dummy hawk on a pole which spun around on fishing wire. I suspect the birds got used to this non-predator after a while, especially as it would get itself in a terrible tangle in high winds. We did find a local hawk owner who came and surveyed the vineyard, but nothing came of it.

Fruit Flies

Let us start with the nastiest fly of all and the one which can devastate grape bunches if not caught early. A fruit fly called Spotted Winged Drosophila (Suzuki fly). The female lays eggs under the skin of the berry which then hatch. To the eye the berry exhibits a brownish colour, is squishy to the touch and has an exit point. This leads to sourness. The effect on the bunch if not picked out can be ruinous to the crop. In 2023, we hear from the wine maker that many vineyards are infested. Why? Maybe a perfect storm with the warm and wet conditions experienced over the summer. The best way to manage this potential crisis is to try and kill the males before they mate while they inhabit the hedgerows during the winter. Not easy as we do not use insecticides as an organic vineyard. There are some organic remedies coming on the market but for the moment the best method is to hang traps out early in the season to catch as many as possible before they lay their eggs. We missed the trick on this in 2023 and had to wade through the vineyard every day for 2 weeks picking out the infected grapes.

Other pests

We do not have any issues with wasps and in fact there is plenty of research to suggest that they do their bit to eat some of the nasty insects. They like to feed on dandelions. Light brown apple moth, European grape berry moth, grape mite, scale insects, capsids, thrips, leaf hoppers, black vine weevils and spider mites can all cause some limited damage leading to botrytis, as well as acting as vector for other pests and diseases. If a vineyard is healthy and has plenty of biodiversity one is going a long way towards avoiding these diseases and pests.

Diseases

Mildews

The UK is a cool climate wine-producing territory and damp conditions are part and parcel of our climate. We must work with it and manage the canopy to prevent mildew taking hold.

Downy Mildew

This can cause total crop loss if allowed to spread. Once established it is hard to get rid of it. Wet weather and temperatures above 10°C cause germination of the zoospores. Symptoms include yellow patches or spots on the upper leaf surface, with white fungal growth on the underside of the leaf. The disease can lead to severe defoliation and yield loss through reduced leaf area. If it spreads to the developing fruit, it produces white furry mycelium causing the berries to dry out. One to avoid.

Powdery Mildew

This appears during hot, steamy conditions. Initial symptoms show as chlorotic spots on the upper surface of the leaf eventually producing greyish-white fungal growth. Severe infection can lead to leaf distortion and gradual desiccation, spreading to the fruit itself in certain conditions. We use sulphur, plus a few biodynamic sprays to counter this.

Botrytis (Botrytis cinerea)

This grey mould is a major limiting factor in UK grape production. The fungus can infect stems, leaves and flowers and, in certain growing conditions late in the season, can spread rapidly. Not to be confused with 'noble rot' – for the production of sweet wines, such as Sauternes, in the south of France. Infections occur primarily following bird damage, hail or strong winds – the skin is split allowing disease to enter the wounds. Disease thrives at 15-20 °C and high relative humidity. Berries are most at risk after veraison (ripening) as higher sugar levels encourage disease to spread. Varieties with thin skins and tightly packed bunches are most at risk. These include the varieties I grow, i.e., Chardonnay and Pinot Noir – especially our Chardonnay. Crop hygiene is the key to tackling botrytis, with good pruning, regular canopy management, leaf stripping around the bunches, and removing infected berries immediately one spots them. Daily patrols late on in the season are essential.

Phomopsis

Phomopsis viticola overwinters in the bud, bark and canes of the infected vines. It presents with pale leaf spots, cracks and girding on the shoots and bleached canes. A wet winter can bring on phomopsis. In the spring the cane shows as silvery often with dieback on the tip and later on in the summer once leaf has appeared leafspot appears as tiny dark spots with yellowish margins on leaf blades and veins. The good news is that it can be controlled to a large extent by spraying sulphur and copper when shoots are around 15 cm.

Dead Arm Disease

The name for dieback on the cane. These symptoms do not usually appear until the vines are at least 6 years old. They are most evident in the spring, occasionally found in a cane one wishes to tie down, when healthy shoots are 20-40cm long. Shoots arising from infected wood appear stunted with small, chlorotic, distorted leaves. These then become necrotic as the season progresses.

Trunk Disease

This is not yet a common problem in UK vineyards. However, if left unchecked the disease can spread rapidly to healthy vines. Young vines are particularly susceptible when the infection moves down the trunk to the graft junction. Infection is from pruning wounds, typically at the retained spur with a Guyot pruning.

2015 April. Field ablaze with bougies

2016 October. Bird scarer does its job

2016 October. Kite flying

2017. Robert connecting wires

2017 April. Heated wire close up with cane wrapped around

2017. Robert trench digging

2023 October. Fruit fly damage

2020 May. Devastating frost

2023 April. Whitehall Vineyard, Wiltshire

2023 April. Birdnetting with Ludo on patrol

Chapter 6

The Art of Wine Making

It took a few years for me to realise that three very different skill sets are required to be a successful wine producer. It is highly unlikely that any one individual possesses all three. You are fortunate to possess two!

One. You need to be able to grow the grapes. If you do not possess green fingers, you will struggle. The following attributes are prerequisites: handling a set of secateurs with confidence, working in all weathers – and I mean all weathers – possessing a strong back (or if not access to a good osteopath or sports masseuse), and good observational skills if not a background in agronomy; or, in Robert's case, plant pathology.

Two. You need to know how to take good grapes and make decent wine. There are a small number of people in the industry who do a double act – that is tend vines and make wine – but they are few and far between. Both disciplines, viticulture and oenology, are taught at Plumpton College, the centre of wine excellence in the UK, yet the majority of graduates veer towards the relative comfort of the winery over the vineyard. Who can blame them.

Three. Sales and marketing. Unless you plan to drink on a Churchillian scale, once the wine has been made, bottled, labelled, and boxed, you need to find good homes for its consumption, with money changing hands. What are the sales channels, how much you can afford to sell at discount to the trade and wholesalers, and how much can you sell from the cellar door? Can you set up repeat sales via a wine club? It is simply naïve to assume that wine, however good, will sell itself. For many unused to this key aspect of business, it will be a steep learning curve to understand the multi channels to market. It is straightforward to get friends and family to buy a case first off – out of loyalty if nothing else – not so easy to make repeat sales to the same customer. Unless the wine is absolute nectar. That is one of the reasons we established vine lease holders from the inception: friends and family who undertook to take large quantities of our sparkles at heavy discount and duty free.

Before we had even planted Robert and I realised that we needed to find a wine maker with a track record. Neither of us had the skills to make wine, nor the facilities. Building a winery from scratch requires deep pockets, plenty of knowhow, not to mention tonnes of grapes to process.

We visited various winemakers but none ticked the boxes until we beat a path to the acclaimed Emma Rice at Hattingley Valley, the winery owned by Simon Robinson. It was an impressive set up. Remember this was 2012 when the UK wine industry was half the size it is today. Fortunately, Hattingley Valley were prepared to take us on without a vine in the ground! If we had waited until 2015 there would have been no room at the inn.

Emma was ably assisted by Jacob Leadley (now heading up Black Chalk) and it was these two who held our hands for the first years when we were keen but green. Hattingley's modus operandi was primarily that of a negociant, whereby they bought in the majority of their fruit from around the country. In some cases, swaps were made with contract customers for grapes in return for wine. We were never large enough to give Hattingley a percentage of our grape volume – we needed all we grew for our own wine.

They handled our first harvest in 2015, followed by the 2016, 2017, the amazing 2018 and finally the 2019 vintage. We became regular visitors to their winery near Alresford for grape analysis pre-harvest, the delivery of the grapes themselves, the dosage tasting, wine collection and other key moments. The problem was Hattingley Valley expanded fast yet we remained a very small producer. Inevitably, we had no choice but to be part of a large conveyor belt where each stage of the wine making process had to be scheduled according to the winery diary rather than in tune with the requirements of our wine. Eventually we were given a polite yet firm notice to go find another winery.

First panic, then a concerted search for a replacement winemaker who would ideally be on a smaller scale than Hattingley Valley. We found ourselves in the position of having no one to process our 2020 harvest. Not a happy state of affairs. Our prayers were answered when we found, via John Buchan's introduction, Daniel Ham at Offbeat Wines. Daniel had impressive credentials, having started out as a marine biologist he changed horses by re-training at Plumpton to become a wine maker. His first job was at Ridgeview, under the wing of trail blazer Michael Roberts, after which he moved on to Langhams Wine Estate in Dorset as head winemaker. With a few very successful vintages at Langhams under his belt, he decided to start up his own show as Offbeat Wines, based at Downton, near Salisbury. His plan was to make his own wine plus that for a select band of small growers who shared his approach to winemaking. We applied and, after a long interview, were accepted! This was 2020 during the Pandemic when the world had shut down. Everything was done online without any face-to-face contact allowed. Daniel was unusual: not only was he producing wine on a small scale, but he was also a biodynamic maker.

Daniel's approach to winemaking

I quote from his modus operandi:

> *'Unlike many recipe-led contract winemaking providers, Offbeat Wines favours an adaptive, vigilant approach whereby the goal is to produce fault-free, terroir specific wines with as little inputs as possible. We stand by the adage 'great wine is made in the vineyard' and would like to emphasise that the quality of wine we are able to produce is largely dictated by the quality of grapes that go into the press. Although we are happy to intervene and make small additions where required, approaches that have the lowest impact on the wine will be adopted first.*

Only under extreme circumstances and at the consent of the client will chemical intervention be employed.

The winery will be managed in a way as to promote a balanced ecosystem whereby a native microflora community aids the wine production process. That said, we firmly believe that wine faults should be avoided wherever possible and will use water, steam and lots of elbow grease to ensure that winery hygiene is maintained to the highest standard. Similarly, we view wine as a living product and are not able to offer sterile wine bottling facilities (necessary for sweetening a wine before bottling).

In order to gauge how each customer would like their wine produced we have also included a basic questionnaire. This is by no means a legally binding document but will be used as a guide to ensure any winemaking decisions are in line with your own values. If you understand these processes and feel happy to complete the questionnaire, then please do. Alternatively, Daniel will be happy to explain the details surrounding each process and work through the questionnaire at a convenient time.'

What breath of fresh air, and so exciting to have someone with this approach who was prepared to handle our grapes. See Appendix 8 for a diagram outlining his approach versus the conventional methods.

It may seem contradictory, yet a high level of skill and experience is required to be a successful 'low-intervention' wine maker, in the mould of Daniel Ham. A strong background in biology and chemistry is a sine qua non, plus many hours spent in the laboratory analysing test tubes. if one wants to leave Plumpton College with a foundation degree in viticulture and oenology.

A short digression. Two professionals find themselves talking to each other at a drinks party, one an accountant, the other a doctor. When the accountant discovers his interlocutor is a physician of many years' standing, he cannot resist telling him his medical symptoms and asking for an opinion. The elderly doctor scratches his chin and thinks. After a few minutes he delivers his diagnosis and prognosis. The accountant is delighted, thanks him profusely, and gives him his business card. A few weeks pass after which the accountant receives an email from the doctor with a bill for £200. He is enraged. He is a time-charge man to his fingertips. He replies to the effect that the doctor only spent 2 minutes of his time yet charged him £200. The doctor replies, 'You are correct, 2 minutes to diagnose the problem yet backed by 40 years of experience to arrive at the solution.' Value for money?

'Low intervention' does not mean sitting on one's hands and watching the wine do its own thing. Remember, left to its own devices, wine will want to turn to vinegar! We found the difference in wine makers working with large wineries and small ones was very marked. At Offbeat, the winemaker is the owner. Daniel will keep a watchful eye at regular intervals – something unknown at a large winery where the head winemaker may be tasting your wine at dosage trials for the very first time. Hattingley Valley were an exception to this.

It is no different from the law firm analogy. At a large city firm, the partner in charge of the deal may only meet the client face to face for the first time at the completion meeting – with a champagne bottle nicely chilling in an ice bucket. What has she or he seen of the transaction? In many cases, very little or nothing at all. The work will have been done by a junior solicitor or pre-qualified junior. This is the way of modern business. It is for this reason that I always try and deal with people on the other side of a transaction who actually carry out the work. Call me old-fashioned, but it has held me in good stead over the years.

The relationship between the wine producer and the winemaker is such a close and crucial one, it is axiomatic to find someone sympatico – or to use a cliché, someone singing from the same hymn sheet.

Winery equipment

At the outset Daniel suggested that it made sense for us to buy our own fermentation tanks and barrels. This would keep not only his costs down but give us more involvement. Additionally, we bought our own stillages – these are steel cages for storing the wine bottles on their sides (approx. 500 to a cage). This, with his help, we duly did. The second-hand oak barrels were bought from a reliable source in Burgundy in 228 litre and 500 litre sizes, whilst the steel tank with a floating lid was purchased from Vigo in Devon.

Daniel, for his part, has spent his investment on kitting out the interior of the 'shed', no small matter, and in purchasing the wine presses. His pride and joy, the 2 tonne second-hand wooden basket Coquard, is the Rolls Royce of presses. Like farm combine harvesters, they cost a fortune but sit idle for 10 months of the years. Bought in France during Lockdown, it was brought across in pieces by three Frenchmen and then painstakingly reassembled in a corner of the winery. It sits and broods. Albeit Coquards are used throughout the world, Daniel's was one of only 50 models made. It is his pension.

Early estimates of harvest volume

I am going to describe the key events in the wine production year when we interact with the winery, starting with the run up to and the harvest itself. This normally occurs at some stage each October.

It is early July and I get the email from Daniel asking for an estimate of fruit volume. We are asked to take random weight samples of bunches and then extrapolate. This is tricky. Over the years we have fine-tuned this process by developing our own system which seems to be accurate: we have three marked vines in each row – top, middle and bottom – and use these vines each year to take a count and then extrapolate. We may be out by 500-750 kilos but generally it is fairly accurate. Across a large vineyard, the process is not so easy and weight estimates can go badly wrong, the consequence of which can be that the winery is short of capacity for press time and fermentation tanks.

In the wet 2023, bunches were much heavier than in the dry 2022 – the rain plumped them up considerably. Fortunately, against a backcloth of a dearth of second-hand Burgundy barrels on the market (French wineries saving money and keeping them for longer), we managed to obtain extra ones by piggy-backing on a larger order placed by Hugo Stewart at Domaine Hugo. The barrels arrived in the nick of time and we were well pleased. This gives us 80% of fermentation in oak and the remainder in steel tank – a good ratio as you need a vessel for topping up the barrels as a result of ullage during the months before bottling.

Testing and tasting the grapes

In the early years, when we were a bit green, we would take samples down to Emma and Jacob at Hattingley Valley well before they were even close to being ripe. 'A bit vegetal' came back the reply on inspection.

The grapes are tested for sugar and acid levels. We want the sugar levels to be as high as possible whereas the acid levels we like to be acceptably low. Every season is different. We use an instrument called a refractometer to test the sugar levels in samples of berry juice. The reading given is for the Brix scale of potential sugar (PA). The other main scale is Oeschle, as used by Offbeat, for which you test using a hydrometer kit. Remember, the potential alcohol for sparkling wine is usually in the range from 9% to 11% and no more: the adding of sugar in poor years increases the first fermentation levels; and of course, the second in the bottle with a little more sugar will lift the final alcohol levels ideally to between 11.5 –12%. One rarely sees bottles of sparkling wine/champagne any higher – if so, it would throw off the sweet/acid balance with any higher levels.

The second key measure is the acid levels – called the T/A (titratable acidity). The main acids found in grapes are: tartaric and malic, tartaric being the dominant acid in terms of percentage. Over the years we have used an acid testing kit and some 'O' level chemistry to arrive at the result.

I will pass over the vendange itself as I cover this in detail later in the book. However, it is worth mentioning that the build up follows military lines with daily exchanges of texting with Offbeat. The date is decided by the grape ripeness and availability of press time. If the weather on the day is sunny, that is a bonus.

The first harvest is always the Pinot Noir as it reaches ripeness before the Chardonnay. The crates are loaded onto several trailers and driven down on the same day to the winey with extreme care. We stop every now and then to check the straps.

The reception party at the winery are there to meet us, at the ready to unload and weigh the crates. I wait with bated breath for the verdict on the quality of the grapes. We get the nod of approval.

In the wine world, 'doing a vintage' is part and parcel for budding professionals wanting to build up experience, and the more countries one can do this in — be it the old or new world — so much the better. It is hard, back-breaking work. France, New Zealand and Australia are popular destinations. It is a bit like being a pilgrim and undertaking the arduous Camino de Compostela. It is a rite of passage. Make no mistake, there is no shirking. Daniel and Nicola take no prisoners but at the same time it is a happy time with plenty of laughter and evening BBQs for the team. Daniel usually makes beer as ship's rations. At Hattingley Valley it was a veritable tower of babel amongst the temporary workers, with young helpers from France, Spain, Portugal, Australia, South Africa, New Zealand and even South America. As I say, the work is unremitting. Like a perspiring engine stoker keeping the furnace of a steam train topped up with shovels of coal, so the winery deck hands are required to keep the presses filled, emptied, and re-filled with grapes for 12-16 hour per day at peak time. The Coquard basket press is heavy duty work. After the first pressing, the press hands need to get into the basket and fork the grapes back into a fresh mountain for the second (taille) pressing. In between, other presses need cleaning out, the freshly pressed grape juice (the must) poured into holding tanks, the floors wiped, crates washed and stacked. At every stage, the winery book records each process. Cleanliness is close to Godliness.

It is 2023. Our Pinot Noir weighs in at 2.2 tonnes. We await anxiously to see if this can be accommodated in the Coquard, the press of choice, which normally has a capacity of 2 tonnes. They succeed, phew. Daniel and Al, who have the job of filling the press, have clearly been trained as platform guards in Tokyo to squeeze in train passengers at peak times. The plates are lifted down into position. The hydraulic arms are now ready to start the gentle squeeze that will last about 2 hours for the first (cuvee) press. You need to be young and fit to do this day in day out for 4-5 weeks.

We return to Offbeat after a couple of days to pick up the clean, empty crates for the second harvest of the Chardonnay. In the meantime, Daniel will have emailed us the results of the PA and T/A. The cuvee and taille produce slightly different results as expected – the T/A for the second press is usually lower but the sugar levels much the same. In 2022, the acid levels on the cuvee were 14.6, on the taille pressing they were 13.6. In Champagne, the cuvée is the first 2050 litres of juice and considered the best. The premier houses often sell off the taille juice. We will ferment the majority of cuvee and taille of each grape variety in separate oak barrels, thus giving Daniel plenty of flexibility when blending. We will co-ferment a certain quantity.

The picking of the Chardonnay follows the same format: it is almost always seven to ten days after the Pinot harvest.

Doing a vintage is a special time for the team. It requires stamina, strength, endurance, and humour. The hours are long, the pay lowish. The payback is camaraderie, plenty of badinage, the chance to meet each producer as he/she brings in the haul. Lifelong friendships are no doubt made. Daniel's special brew oils the works and keeps the team happy.

A word about the Pinot Noir grape. As we all know, the Pinot is a red skinned grape – actually deep purple. It is normally associated with red still wine. When making brut champagne/ sparkling wine, however, we want just the juice of the pinot which is white and not the skins. Therefore, a gentle pressing is undertaken. Making brut rosé is another process which we will come back to later.

A short digression. Here is an interesting story behind the acquisition of our picking crates back in 2016. Through our wine forum I connected with a vineyard down near Exeter which had equipment for sale. The owner, it transpired, had produced 10 vintages and was now selling up. He was an intellectual property lawyer by training. He told me that he had been advised by the instructed estate agent to grub up the vineyard to make for an easier sale. The house itself was lovely and so was its position. He followed their advice and sold the property. Sometime later we heard through the grapevine (ha-ha) that the new owners had replanted. The moral of the story – be careful when listening to the advice of agents as they are simply acting for the deal not the seller.

If we ever have to sell Orpenham on the grounds of being too decrepit – I pray I will be carried out in a box – there will be someone out there, a minority I accept, for whom the vineyard will be a big magnet. With under two acres, it will be a doddle to look after and drink the proceeds! Some houses sell with a vineyard planted but not yet in production. Clearly this will be a drain on resources for years to come but for new owners the heart reigns over the head.

The pressing and first fermentation

Daniel believes firmly that 80% of the work is done in the vineyard, 20% in the winery. This may seem a truism perhaps but one just cannot conjure up good wine from sub-standard grapes.

From the moment we moved to Offbeat and adopted organic principles in our viticulture, the fruit quality of the grapes improved markedly, and the sweetness levels rose in tandem. In most years, Daniel will expect to add no sugar in the first fermentation – called chaptalisation (named after Jean Antoine Chaptal, a French chemist) – nor cultured yeasts. There will be just a spontaneous fermentation of bubbling fruit. In a difficult year where the potential alcohol is low, he will add just a small amount of sugar. He is not rigid in his thinking and believes that sparkling wine needs to hit certain alcohol levels – or otherwise tastes 'thin'.

Malolactic fermentation

This is the process whereby, through bacterial change, the malic acids are converted to the softer lactic acids (as found in milk). This process can happen automatically but if not then the winemaker will intervene. In England, where we have high acids, this approach is almost uniform. Interestingly in Champagne where temperatures have been steadily rising bringing unwelcomed high levels of sugars and acids, the winemakers will commonly intervene to stop malolactic fermentation happening. Take the great, hot summer of 2018 when England and Wales enjoyed a fabulous harvest of nigh perfect fruit; the same was not the case across the Channel where growers worried that the acids would fall through the floor, making the wine lose the necessary tension between sugar and acid.

November and December

As we get into November, temperatures in the winery are getting decidedly parky. The first fermentations should be generally finished so that the wines can sit out the cold period in a stable condition. At this stage the winery will be spotless, the barrels all marked and stacked neatly, and the doors firmly closed to the elements. Sometimes it is too cold to come into the winery – Daniel and Nicola will work from their desk at home. There is plenty of paperwork to do – ordering all the supplies for the following year and chatting to their contract clients.

February onwards
This is the time of year when things start up again at Offbeat – disgorging, adding dosage, corking, foiling, labelling and despatching wine, both for themselves and their contract clients. If they are doing a run of finishing wine for us, I will drop down with supplies of wireheads and labels.

As I write, there will be no fewer than 5 of our vintages sitting at Offbeat in various stages of production. This is a quite typical amount when producing sparkling – unlike still wine. Daniel keeps meticulous records of all the wine he makes for himself and others. Not only that, but he will also be keeping an eye on each barrel and tank.

August – Tirage (bottling)
It is time to start the second fermentation in the bottle, where the wine

is drawn from the fermentation barrel and poured into bottles with the addition of a little sugar and yeast to ensure a successful fermentation. Before this can take place, winemaker and producer must sit down and agree on what styles and blends they should make. Some nine months after pressing, usually around June each year, we go down to Offbeat and gather around our barrels with glasses and spittoon at the ready to have our first taste of the base wine. We all have notebooks to write down our marks and impressions of each barrel sample.

The 2022 wines are startling for the purity of the fruit – I have difficulty spitting out and just roll the wine around in my mouth and let some slip down the throat. A portion of the Chardonnay and Pinot has been co-fermented in a steel tank – that will become the basis for the brut blend. All the barrel base wine is ready, with the exception of one which needs longer in oak.

We decide with ease on the base wine which will be used for the brut reserve, the brut rosé (again a 100% pinot with some reserve red wine from the year before which is sitting in a spare container). Then, lastly, we taste the best barrel of Chardonnay. Should we make a still Chardonnay? Daniel thinks it would make a fine still wine at about 11%, with regular battonage of the lees and leaving it in barrel for 2 years. We ruminate. I have always wanted to make a blanc de blancs (100% Chardonnay) and I believe this is the year in which to make a limited quantity. One 228 litre barrel will make about 300 bottles. After looking at the two options, we plump for making the BdB. It will be a long wait, yet a wait worth waiting for. We are thrilled at the prospect. When another good, dry, hot year materialises – wet 2023 is not the year – we may try and make a still Chardonnay.

September

A crown cap (like a beer cap) is affixed at bottling. The cork comes later. The empty bottles arrive pre-sterilised and are then stacked onto the production line before they are filled with wine. There is no filtering. On a good day, without any machine breakdowns, Daniel and Nicola can handle, without interruptions approx. 3000 bottles per day. Once filled and closed with a crown cap, the bottles are laid on their sides in each stillage (wire cage) in a strict pattern.

The bottling line is set up and Daniel and Nicola set about preparing the agreed styles and blends. The bottles will go into stillages sitting on their side where they will 'sleep' for a minimum of 24 months in the cool, darkness of the storeroom. For the 2022 brut reserve and brut rosé, this means disgorging no earlier than August 2025 – although having said that the rosé can be ready a little earlier.

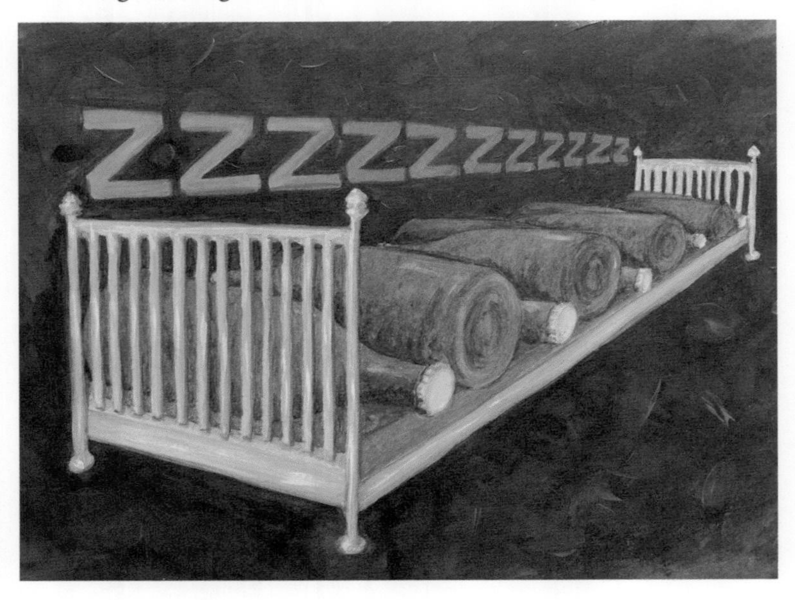

Riddling, disgorging and dosage

24 months down the line, the next stage is to extract the dead yeast cleanly from the bottom of each bottle. This is performed by riddling, a process whereby, in modern times, the stillage is placed on a gyropallette. In the olden days this would have been performed by hand in a pupitre (thanks to widow Madame Clicquot).

It is always a joint decision as to when the time is right to disgorge – finishing the wine off by riddling the bottles, removing the dead lees (yeast) and adding a small quantity of liqueur de dosage (sometimes referred to as liqueur d'expédition in champagne).

Before this final process can be started there has to be agreement of the level of dosage to add. Daniel lays out six 'blind' bottles with different dosage levels. We will discuss dosage levels in Appendix 4.

The beauty of being with Offbeat is that we are not obliged to process the whole vintage in one fell swoop as was the case at Hattingley. Indeed, Offbeat, given their size, would struggle to do this. Instead, we will riddle maybe 500-1000 bottles at a time. This is a syrup mix of our wine and a small amount of sugar. After disgorging and dosage, the wine will then need to sit for a

further 6 months 'on cork' to settle after being disturbed. Around November 2025 the foil for the 2022 vintage will be slid on over the wire hood, and three labels attached. Offbeat is a bonded winery which means the duty must be paid before they can ship to us. That is always a nasty, nasty invoice to pay. Therefore, by around December, we can take delivery for onward sale to our expectant customers.

Each vintage is quite some journey from grape to bottle, with the endgame of producing sparkling wine to delight our customers, we hope, when they pop the cork. Eyes, nose, mouth – in that order.

Gravity fed wineries

Building one of these from scratch will start at £5m! It involves creating a building with multi levels (a small multi storey car park) in order for the wine to flow downwards by gravitation rather than be pumped around from tank to bottle thus disturbing the wine. The presses sit on the top floor (or can be craned from ground floor to the top), followed by fermentation tanks on the next, then to the ground floor for riddling, disgorging, labelling and despatch.

I have seen two such wineries: one in Otago in South Island, New Zealand at a famous winery called Felton Road, owned by Nigel Greening (events entrepreneur); the other at Ca'n Axartell in Mallorca, owned by Mr Schwartzkopf (a German shampoo billionaire) which is deeply impressive having been built into the side of an old quarry. One can see that it takes big bucks to fund these enterprises. Plus a passion for detail.

Ca'n Axartell in Mallorca

Ca'n Axartell's roof is grassed over and one can see through the large sky lights goats grazing contentedly. I asked Patrick, the owner of the Goat on the Roof in Newbury if this is what gave him the idea to name his restaurant. No, but what fun! Both wineries are impressive, both are owned by men with deep pockets.

At Offbeat, Daniel also believes in no pumping either (nor fining for that matter). His solution is a wee bit less expensive: he uses a forklift to hoist up the barrel to a sufficient height to let the juice flow by gravity. Much care is needed with such a valuable consignment.

Dear reader, now you have read and digested this chapter on wine making, I think you will understand why, at the outset, we were sensible enough not to attempt this process ourselves.

Chardonnay bunches with a few weeks to go

Pinot Noir, groaning to be picked

2018 April . Our 2018 wine in cages

2020. Red wine fermenting

Daniel on forklift duties for our crates

2020. Our winemaker, Daniel Ham

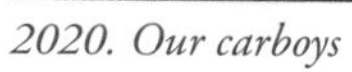

2020. Our carboys

2020. Pinot Noir in the Coquard

2021 Oct. Daniel works the basket press

2021 December. Daniel plus Coquard

2023. Tanks and forklift at the ready

2023 October. Small basket press and tank

2022 June. Tasting 2022 wines

2023 March. Our 'demo' pupitre

2023. Bucher press

2023 October. Chardonnay in the press

2023 October. Chardonnay cake

Chapter Seven

The Road to Damascus

Our Conversion to Biodynamic Viticulture

Ray Hamblin, the farmer who lived at Orpenham Farm before us, was born in the farmhouse in the late 1920s. He had a full life in many ways outside farming – cricket, reading, game shooting, ploughing competition judge, owner of ancient tractors, pheasant plucker par excellence – yet, over 50 years of running a mixed farm of 250 acres, he did no more than scratch a living. His father and grandfather had farmed the same tenanted land before Ray with the same results. It was a way of life. Ray retired in 1979.

Over many a conversation with Ray – one needed to put aside at least an hour as he did like to talk – he would reminisce about the changes in farming that he had witnessed over his lifetime. The dramatic increase in cereal yields through industrial-scale farming was just one of the main topics. With the introduction of satellite aids in the cab for drilling, the drill could triple seed on a barren patch. Together with high levels of inorganic fertilisers spread on the soil, the estate managed to obtain triple the yields over what Ray has achieved on the same fields. But at what cost to the soil? I suspect that the only thing today that sustains crop growth in the fields that surround Orpenham Farm is water.

During the Pandemic I read two books that had a considerable effect on my thinking about soil in general and the vineyard in particular. The first was called *Dirt to Soil* by Gabe Brown, a prairie farmer from the US; the second was *English Pastoral* by James Rebanks (author of *The Shepherd's Life*), an elegy to farming in the Northern Fells where he inherits a family holding and tries to revive a livelihood, based on working the land in the old way as taught by his grandfather – something that had vanished from the landscape.

Let's consider Gabe Brown's book in more detail. In the opening couple of pages, he sets out his stall. It chimed with me when I first read it and subsequently, once we had decided to go organic in the vineyard, it just made such a lot of sense. 'Our lives depend on soil,' Gabe states. His five principles of soil health are laid before the reader.

1. Limited disturbance of the soil, be it mechanical, chemical or physical. Tillage destroys soil structure.
2. Armour. Keep the soil covered at all times.
3. Diversity. Strive for diversity of both plant and animal species.
4. Living roots. Maintain a living root in the soil for as long as possible during the year.
5. Integrated animals. Nature does not function without animals.

He took over a dust bowl prairie farm in the US northwest and transformed it, through regenerative farming, into a profitable and healthy family farm. He has become, according to the book's blurb, 'the voice and face of regenerative agriculture around the globe, inspiring a movement that is reshaping the future of agriculture and the way farmers, consumers, and policy makers think about sustainability.'

I concede that 100% organic farming is not going to feed the billions on our planet. However, consider the huge amounts of food we throw out from our fridges uneaten; plus, the low levels of nutrition available in much of the industrially produced food we consume every day. Vegetables bought in a supermarket may look sleek and uniform but rarely do they contain any goodness. Take the humble carrot. The carrot of 40 years ago might have looked wonky but it contained 20 times the nutritional value of its 2020s' successor. Maybe, muses Brown, that is why Americans need to take in such huge portions! As he travels around America giving lectures, he notices wherever he touches down, every hotel he stays in serves up the same homogenous and tasteless food as the last. It adds up.

Why attempt to produce organic wine in England? Surely the damp climate and the concomitant mildew issues makes it a tough challenge. Hugo Stewart, of Domaine Hugo in Wiltshire, successfully ran a biodynamic vineyard in the southwest of France for 20 years. He refutes this, claiming that the warm, humid nights one experiences in the southwest of France are more conducive to mildew than in England. He has planted vines on his farm in Downton, just outside Salisbury, and runs it on biodynamic lines. He has had great results with his English sparkling wine. He firmly believes that one can use biodynamic viticulture methods effectively in a climate as damp as ours where the biggest foe is mildew. It is mostly timing, he says, as to when one applies sprays.

There is much talk about 'sustainable' viticulture in English wine circles but what does it actually mean? Is it just virtue signalling? As Alan Chubb of Quoins Vineyard in Somerset remarks: "I have concerns about the public confusion over 'Organic' and 'Sustainable' when there are some obvious differences. Being organic is not using ANY unnatural chemicals as opposed to using artificial chemicals when needs really must. I would underline here that if you are using artificial chemicals even if on a 'needs-must basis' then you are doing something wrong or not addressing the real cause of your problem."

The big step up from 'sustainable' is organic which cuts out spraying nasty stuff on the vines. Adding plenty of organic matter from grazing sheep is another plus. Yet, I felt that there was something missing in organic farming. It is the regenerative approach found in biodynamic principles that addresses the soil itself and the manner in which one can improve its microbial life.

Another of the attractions of becoming biodynamic is its holistic approach to the farm and the biodiversity around it. Instead of avoiding just fungicides as with organics, biodynamic farming uses, as devised by Rudolf Steiner, a series of applications which improve both the soil and the plant. This builds up the vine's resistance to disease. Using the antibiotic analogy, if you use too much the drug will become ineffective (as has been found with superbugs). It is no different from planting a vine and then plastering it with synthetic chemicals constantly. They simply stop working.

When we moved wineries from Hattingley Valley to Offbeat in 2019, or I should say when we passed the online interview with Daniel Ham during the Pandemic, I began to think a little bit more about organic wine growing. Daniel was an impressive advocate of regenerative farming and low-intervention wine making. His background was in marine biology before retraining as a wine maker. We also got to talk to Hugo Stewart on whose farm Daniel had built his winery. Hugo came to see us at Winding Wood and thought there was no reason why our site could not become biodynamic. 'You have plenty of wind here,' he chortled, 'that is a good start!'

The paddocks on which we planted our vineyard have never, as far I know, been ploughed for arable crops nor been sprayed with any synthetic sprays. Hundreds of years of sheep and cattle grazing has created a fertile topsoil with plenty of flint and chalk below. The Pinot field was once an apple orchard, according to Ray. That was until Ray's father allowed a drover to park his sheep one night in the field on his way to market. They completely destroyed his apple trees.

In 2021, I told John Buchan, our agronomist, that we planned to grow our hair long and go organic, so please would he devise an organic spray programme. He did not demur although there was a sharp intake of breath. He produced a programme which included garlic, seaweed, sulphur and small amounts of copper. I was mighty pleased to sell off to John the contents of our chemical cupboard.

Robert, my partner in crime, had been a farmer and dentist, yet he was more sceptical than I that without the full armoury of artificial fungicides and herbicides at our disposal, we had any hope of protecting the vines in a wet season. I reasoned with Robert that there were plenty of reasons to convert: Daniel made biodynamic wines and if we had surplus grapes in the future, he would only take them if they were organic. Furthermore, the market for English wine appeared to be becoming crowded with the relentless frenzy for planting by the uber rich. Would we be able to find a market for our wine with all this competition. Organic wine would give us that USP. And then there was the environmental consideration of pouring chemicals on our fruit. Anyway, said Daniel, he was not convinced that chemicals do much good at arresting disease.

Biodynamics was a philosophy promulgated by Rudolf Steiner in the 1920s in response to what he saw as the impoverishment of agriculture by the use of artificial fertilisers. He devised a system with a focus on creating optimal soil, plant and animal health, providing balanced nutrition and supporting health immunity. He argued that when farms incorporated a robust diversity of plants and animals and created a habitat for natural predators, pests and diseases then had few places in which to thrive.

Fundamental to biodynamics is its 'salutogenic' approach to the farm/vineyard: in others words the primary focus is brought to bear is on achieving health instead of fighting disease.

We started organic viticulture in the season of 2021 – a difficult year for all sorts of climatic reasons. By 2022, our second year of being organic, I was introduced via Daniel Ham to David Morris who, after his first visit, then became our biodynamic consultant. It was he who gave us the confidence to start using biodynamic applications and sprays. No, we did not grow our hair and start smoking weed. Biodynamics has something of a hippy-dippy reputation but there are plenty of advocates who do not grow a ponytail and wear open-toed sandals.

Dave arrived at Winding Wood for his first visit on a cold day in February. He had driven all the way from Monmouth where he is based. The first thing he needed was a strong coffee. He said we were just about at the maximum of his range. He produced a miniature notebook and pencil from his pocket and then, at my suggestion, took a walk through the rows, inspecting the vines. He seemed shy and perhaps was asking himself if we were serious clients. I assured him that we were committed to organics but would need considerable hand holding. I hope he was at least impressed that all the work in the vineyard was done by the owners, bar some of the repetitive tasks which we passed to our Romanians. He took plenty of notes and followed this up next day by sending a thoughtful and thorough report. He was the man for us.

The vines in 2022 had never looked healthier. Even during the driest period of the summer our canopy looked amazingly green and healthy. The roots of the vines would be probably 6 - 7 feet deep by now, enabling them to reach water as they needed it.

In addition to producing good fruit, we also want to improve and regenerate the soil. We have never had intensive agriculture on our land fortunately – just centuries of sheep – so we had a head start. Nevertheless, it is crucial to keep improving the structure of the soil.

At a 'Three Wine Men' event in London recently with Oz Clarke, one of the UK's leading wine writers, he talked about the famous Chapoutier vineyards in the Rhone where they are fully biodynamic. The owners quite seriously say that 'soil talks if you listen to it.'

One square metre of healthy soil will contain 5,000 insects, molluscs, and worms. It will also contain 300 different yeasts. Compare that to the equivalent soil in an agri/industrial field: there will be no yeasts, no ants, and no worms. Good soil amazingly replaces itself completely on a regular basis. Therefore, one of the major things we are trying to do is improve microbial life in the soil. Do that and the vines will prosper of their own accord.

What sprays do we use? In addition to using sulphur and a tiny bit of copper, we apply seaweed and garlic. To these we have added other biodynamic applications such as horn manure (for the soil), silica, yarrow, willow bark (it contains salicylic acid as found in aspirin), and horsetail. Most of these are aimed at building resistance in the vines against mildew. These are all plants with medicinal properties.

Performing certain tasks in the vineyard and spraying the canopy according to the phases of the moon may sound cranky but if one stands backs and thinks about it, maybe it is not quite so nutty. After all, medieval farmers would have worked around these cycles, knowing when best to work the soil and plough. As the earth spins, different areas of the planet face the moon, and this rotation causes tides to circle around the planet. This movement of water is crucial. There are times to plant, times to prune (when the sap is rising and thus heals the wound), times to harvest, times to do various tasks in the vineyard, and of course, times to do nothing but rest. And remember that we humans are 60% water!

When it comes to viticulture, my goal is always to produce the best grapes that one can. It is impractical to follow everything that Steiner dictated to the letter at the expense of the wine. There are times when one just has to get out and spray the vineyard because wet weather is forecast but the moon is not in the right place. I have been to vineyards where good environmental practice of the land seems to take precedence over the quality of the wine produced. You can allow vines to grow all over the show but not at the expense of high-quality fruit. I personally do not like cloudy wine or pet nats (petillant-naturels) at 8 % alcohol but many young drinkers do. It is a matter of taste.

For the 2022 season, Dave Morris designs us a new programme whereby the biodynamic applications and organic sprays both run in tandem. Tony Egerton will run the main air-assisted sprayer for sulphur and copper (cab needed for this) and I will use the pull behind sprayer hooked up to my Raptor (ride-on mower). While 2021 was a light year in terms of yield following on from the disastrous 2020 when we were badly hit by an unseasonal late May frost, thankfully 2022 turns out to be a magnificently hot and dry summer with temperatures hitting 40°C in July. There is next to no disease to worry about. The quality of the grapes was outstanding and Daniel was over the moon. 'Now', he said, 'I can without doubt make some great wines with this fruit.'

2023 season. This was as testing a summer as I can recall in ten years. The biodynamics helped enormously to fight the mildew if not eradicate it completely. I placed a rain gauge in the vineyard to record levels with alarm bells ringing if we had more than 10ml in any 24 hours. This occurred four times in July – unprecedented. There were many a frantic text to Dave. Getting on a spray within 24 hours of a heavy rainfall became nigh impossible as the vine canopy had barely dried out before the next deluge.

If not for the superlative June with first class flowering, the season would have turned out to be a disaster. In the end we harvested close to 5 tonnes between the Chardonnay and the Pinot. Not a bad harvest.

During the summer I mastered the biodynamic applications – the teas, the sprays – and moreover, I knew where to source ingredients like young willow whips, yarrow and the dreaded horsetail. The towpath of the Kennet and Avon canal at Kintbury provided the willow whips in profusion. The last application of the season, after the harvest, was the horn manure (B500) which we mixed/ dynamised for 1 hour in a large tub and then cast with a large paint brush with big dollops. This is an accelerator for improving the soil life. There is a fascinating and complex world just below our feet.

As I mentioned above, we use plants with medicinal properties out of which we make 'teas' for spraying during the rainy July. The sprays are either standard organic or biodynamic: we use sulphur and a tiny bit of copper plus chipped, young willow whips.

The willow shoots are fed into a garden chipper that neatly chops them into small pieces. We then simmer in water for two hours after which it is left to steep. The concentrated solution is then diluted with water and added to the spray tank. We also use dried horsetail that invidious 'weed' to gardeners, which is simmered in water and then steeped for 2 hours. Horsetail is full of silica whilst willow bark contains plenty of salicylic acid.

The art of dynamising . A dynamiser is a round container filled with water – we use a barrel or large terracotta pot – to which the preparation is added slowly through the fingers. Quoting Britt and Per Karlsson in their book, "The entire contents are stirred, first in one direction, then quickly changing to the other. This agitation creates a 'vortex', and the change of direction creates a 'chaos'. Water is considered in biodynamics to have 'memory' and so it will memorise the power of the biodynamic preparation and then pass it on to the vineyard."

We are in transition to becoming certified via The British Biodynamic Association as an organic and biodynamic vineyard, hopefully by the year end. It takes 3 years, with annual inspections, the submission of our spray programmes, and even the odd spot check. In a wet year we need to obtain a derogation to use sulphur and copper in higher-than-normal quantities. It is a relatively expensive route, but well worth gaining the Demeter standard.

There are only a handful of vineyards in England and Wales which can claim to be biodynamic, and only a tiny number of those produce sparkling wine. From a total of 1000 vineyards and maybe as few as 30 are commercially organic or biodynamic.

Although we do the large proportion of the viticulture work ourselves, we do lean on Ed of Mitcham Viticulture to lend us one of his small gangs run by the hugely experienced Doru. There are various tedious jobs which they can do in a day that would take me 10 days such as tying down vine canes. Then, later in the growing season, we give them the task of removing early laterals in the fruiting zones, bud rubbing, and leaf removal from around the bunches in July/August. The winter pruning is too important a job to be left to them alone as they will never have the time to be as meticulous as us. It is by far the most important task to be done in a vineyard, make no mistake.

The larger the vineyard, the more reliant the owner is on contract labour. There are three main suppliers in the country: two based in Sussex and one in Essex. The fourth, run by Ed Micham, as mentioned above, which operates from a base in Oxfordshire, is on a smaller scale than the others. The proximity of his workers makes life a lot easier for travelling to Winding Wood.

We have now established a triumvirate of expertise to call upon: Daniel in the winery (he is also a first-class pruner), Dave on the biodynamics (he also is a winemaker), and Ed to provide manpower. They are all either in their early 40s or approaching that milestone, so years of working experience between them. These wingmen are each brilliant in their field and we are thankful for their help. This seems a wise move for a pair of owners now in their late 60s. Inevitably, it is a case of *anno domini*. Backs, wrists, knees – they are all taking a pasting with working at tricky angles and heights. One needs a good osteopath (sadly retiring) or sports masseuse (a brilliant find locally) available at the drop of a hat or a pair of secateurs. In my case I have even had treatment from an orthopaedic surgeon for hand issues. But for arthritic joints they can do nothing – except new fingers. That might be extreme.

Dave and I always have a chuckle about the potential vineyard owners we meet along the way and their mixture of naivety and enthusiasm.

He always asks a potential planter a series of questions along these lines:

1. Do you have any idea how much work is involved in planting and working a vineyard?
2. Who is going to do the work?
3. Who eventually is going to make your wine?
4. Where are you going to sell the wine 3-5 years down the road?

Dave has one big project on the go near Monmouth which does stack up. Several landowners with adjacent land are going to club together and plant simultaneously, share the expensive equipment, such as tractors and sprayers; and, of course, will help each other with the viticulture. This, Dave says, is a well-thought-out plan for cooperative viticulture.

A short digression. The interview and the hand inspection. Back in the 1990s a certain Mike Bayon ran a garden company in Southwest London. He was a true eccentric. Mike had been a front gunner in Lancaster bombers during the WW II at the tender age of 20 and came through the experience unscathed. After university, he became a classics teacher at St Pauls before eventually deciding that he preferred gardening to teaching boys. Whenever he interviewed candidates for garden maintenance – mostly Antipodeans – he would always ask them to show him their hands. If they were not scarred by calluses, a mark of having used a spade and fork in earnest, they failed to get the position.

The vineyard world in England is divided between those who know how to handle a pair of secateurs and those who only pick up a pen. Are you a man of the soil or just someone who employs others to toil on your behalf and have no real idea of what happens in the vineyard?

Robert used to joke that he was the blue-collar worker and I was the white collar in the partnership, ie doing the paperwork, the marketing plans, the accounts etc. He liked nothing better than to get his hands dirty and was happiest under a tractor engine with a spanner in hand. In truth, we did many of the vineyard jobs in tandem – exchanging views on each vine as we went along the rows. In the early days, Robert was usually the surgeon and I was his apprentice holding the tools. As a former dental surgeon, Robert would say that addressing a vine for winter pruning was no different from approaching the mouth of a patient in the chair: always plan what the result should look like before picking up the drill. Winter pruning a vine is no different: there is no point making the first cut before one has decided what the final balance and shape will be. Good pruning will ensure a healthy vine.

As I hope I have made clear, careful pruning is key to ensuring the vine's health, managing the energy burst during the season, keeping it free of disease, and making sure that there is balance.

2023 August. Horsetail and Willow bark

2022 June. Christopher stirring the BD500

Leonie takes over stirring after half an hour

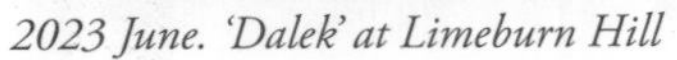

2023 June. 'Dalek' at Limeburn Hill

2023 August. BD sprayer with jets on

Chapter Eight

Diaries, The Later Years

(2022-23)

Here follows extracts from my vineyard diary for the 2022 and 2023 seasons. Weather-wise – two very contrasting seasons.

A Brilliantly Hot Season

April 2022

Having evicted our sheep in 2012 in order to plant a vineyard, we had a reunion of sorts in February when we invited a few ewes back for a mini break to do a bit of grass nibbling, courtesy of local farmer Trevor Gore. Trevor was silently a bit worried that they might start chewing the trunks of the vines, as can happen with young apple trees, but they stuck to the main task and left the vines alone. They also provided plenty of natural fertiliser.

Our organic journey continues as we enter our second year of 'conversion' (of the non-religious sort). In February, we planted a wild-flower bank below the Pinot field which is bedded on membrane to help get it established. This should look terrific come the summer.

The gruelling task of winter pruning is complete, and the new vine canes are tied down tightly on to the heated wires. This frost protection system has served us well even though the wires have been cut with secateurs on occasion (yours truly guilty). The next month is always a nervous period when the tender buds are vulnerable to spring frosts. To date, the buds are dormant after mixed weather in March and the beginning of April.

The team of Romanians led by Doru help me tie down the vines and will then help again in the summer. There is nothing the incomparable Doru does not know about caring for vines.

We have just released our first sparkling rosé to add to our range so we now have three styles of wine for sale: brut, demi-sec and brut rosé. Coral pink in colour with lots of red fruit, our brut rosé should be a winner this summer. All the wines from our 2018 harvest reflect the wonderful summer we had (the best in a generation it is said) with perfectly ripe fruit. This vintage has our lowest dosage to date.

May 2022
11 May, we have '3 leaves out' and all is looking even and healthy. We suffered a severe night frost on 25 April and the heated wires came on. More damage to the Chardonnay than the Pinot for some reason.

June 2022
Flower power: good flowering shown here for the Pinot at the end of June, and all bodes well if the temperatures stay warm. What we want is fast and consistent flowering.

July 2022
'It ain't half hot Mum!' (Showing my age here.) The gauge showed 40°C degrees in the shade. This must be a record. Flowering went very well. We now have fruit set at which stage we will decide how much leaf to take off. In hot climate wine producing countries, it will be the reverse: shade the bunches with as much leaf as possible. We recorded 41°C degrees in the shade on the day when the Met Office instructed us to stay inside and pull the curtains! Nanny State. I was fishing on The Tweed – we had to abandon the river and sit inside the cottage with the curtains closed.

August 2022
The rain finally came this week after 6 weeks of drought and high temperatures. When the heavens finally opened it was a blessing

for the grapes, even though the canopy had remained remarkably green and healthy throughout the hot weather. The vine roots will be at least 6 feet deep and be able to reach through the chalk to find water. If we had planted this year, I might not be so relaxed.

Flowering started in the Pinot field at the end of June and went through both varietals without a hitch. This was followed by fruit set with nice balanced berries forming as bunches. The wine maker, Daniel, has sent me a harvest estimate matrix to complete. This has me scratching my head as it is very difficult at this stage to estimate what the final weight will be. It will be lighter for sure than in other years due to the hot weather; however, if we get plenty of rain at veraison to plump the bunches up, who knows. The difference between an average of 120gm per bunch and 100gm will make a difference in a tonne either way (the equivalent of 800 bottles). Then the question is, do we have enough oak barrels to ferment in? Ordering from Burgundy is not as easy as it once was, post Brexit, now with a lead time of 6 months.

We entered a couple of UK wine competitions this year and managed to come away with one gold and four silver medals for our range of 2018 fizz which was rewarding. Nice to get a gold medal for the first release of our sparkling rosé. It has been selling like hot cakes this summer. We made it by blending our 2018 brut with the ripest of the 2022 Pinot Noir to achieve the colour and taste we wanted. There is more delicious, pink, foaming wine in production to be released early next year.

September 2022

If you look carefully, you will see some blue bird netting on the outside row of the Chardonnay field. As the bunches get riper, birds such as blackbirds start to peck away. This year we are trialling netting to the outside rows in a bid to deter them. The weather during September has been mixed: sunshine and showers. This is always a month when we worry that the grapes are not going to ripen sufficiently with the days becoming shorter –fewer growing degree days. This is usually unfounded unless a poor September is followed by a wet October. Veraison started well and there is good balance in the berry size. This is the month we undertake green harvesting in earnest – removing bunches that are well behind the majority – and this is a task we will do each week.

October 2022

We pressed the button early, in the middle of October: the grapes came off in a total of 6 hours after 7 months of close attention and nurture. I then promptly went on vacation to let my back recover, feeling relieved but a bit flat.

If 2021 qualified, following a thoroughly wet summer with a low crop, for what our late queen famously coined (after the fire at Windsor Castle), an annus horribilis, then this year will go down as an annus mirabilis. We were owed one.

We had a beautiful crop of grapes – not an enormous tonnage as in 2018 (that was a season in a generation) – but the quality was magnificently ripe and disease free. The six weeks of very hot weather in July/August probably contributed to smaller bunches; furthermore, we were growing fruit on last year's cane which had not been very productive. The canes in the Chardonnay block had plenty of 'blind' buds. On top of that the roe deer dawn breakfasted on a few young buds during the spring. They are becoming quite a pest. Another result of the Pandemic and low levels of culling.

Back to the winery at Downton. The first fermentation of our juice was smooth and instantaneous in the warm conditions of October, with the winery doors flung open. High sugars and wild yeasts made it a stress-free period for Daniel.

Before the harvest itself we had done a month of 'green' harvesting. This involves taking off any bunches that are unlikely to ripen, reducing bunches on some shoots, or those with a bit of rot (often where a bird has taken a peck out of a berry). This makes harvesting itself a doddle so that pickers can just snip, snip. In previous years we had asked the pickers to do surgery 'on the go'. This has meant slow progress and a very tired workforce.

The total tonnage was just under 3.5 tonnes which was not large but as predicted. The Pinot produced more than the Chardonnay which was unusual – largely down to the fact that the

Chardonnay fared worse last year, thus affecting this year's crop.

We will look at what styles to make early next year but given it was such a good year for Chardonnay we may make a blanc de blancs – 100% white grapes, normally Chardonnay but not exclusively. Breaky Bottom in Sussex produce a fine blanc de blancs with 100% Seyval Blanc. Peter Hall is something of a wizard.

I have one more biodynamic application to make before the vines go to sleep for the winter: horn manure mixed with a compost tea which will be sprayed on the soil before temperatures drop too low, i.e., 10°C.

November 2022
Daniel siphons samples from our barrels of the 2022 wine for us to taste. These four oak barrels hold some of the cuvée and the taille (the second pressing) from both our Chardonnay and Pinot Noir. This allows us options for future blending. This season really turned out to be the year for Chardonnay – that trickiest of grapes to grow but when it comes good it is unbeatable. Our current plan is to make a small amount of blanc de blancs (100% Chardonnay). If there is any left, maybe even a still Chardonnay, says Daniel. We have also co-fermented a volume of Chardonnay and Pinot in a steel tank – we were too late in ordering another oak barrel from Burgundy.

When we visited the winery in late November it was immaculately tidy with all tanks full and oak barrels racked three high on frames. A far cry from the orgy of pressing during October. The winery was distinctly chilly, and it will not be long before the wines are tucked up for the winter. Daniel will visit periodically over the next two months but otherwise work from home in Salisbury preparing for the new season.

We took the opportunity to taste some disgorged wine from the 2019, 2020, and 2021 vintages. This is 'naked' wine in the bottle before dosage has been added. The plan is to release some of the 2019 vintage next year. Daniel has made a sparkling rosé from a portion of the 2020 vintage and we all agreed it tasted very promising. Lots of lovely red fruits coming through. This will be ready in the early summer of 2023. Sparkling rosés are often ready for release earlier than their brut brothers thanks to the extra still red wine added in assemblage.

December 2022

Temperatures for the last week have been sub-zero. A hoar frost envelops the vines which lasted throughout the day. Good news for killing any nasty mildew spores hanging around the vineyard. We held our first pre-Christmas 'open' weekend at the vineyard from Friday the 9th to Sunday 11th to which we invited our local customers. As the tasting barn registered -1°C despite space heaters blasting, we decamped to The Dairy across the lawn. This we got nice and toasty. In addition to serving our fizz with mince pies, we assembled a few other local producers of 'artisan' food. These included organic meat from Grove Farm at Stitchcombe, fudge from Marsden Fudge, smoked trout from The Berkshire Trout Farm, honey from Berkshire Bees, gin from Hawkridge, and, last but not least, jams and chutneys from Annie Adde of No Adde-tives.

A steady dribble of people dropped in over the three days some dressed up like Cossacks to ward off the cold – and we met some delightful new people who we hope will return in the summer. Even some visitors from The Retreat at Elcot Park who strode across the fields from the hotel in hired wellies.... and had to return with cars to load up their purchases.

I have spent the last few days packing boxes of wine for delivery locally and for further afield by courier. Always nervous despatching wine to addresses in The Smoke yet with modern tracking it all seems to work seamlessly. The trick is to use 3-part pulp to protect the boxes as 'fragile' does not seem to register with couriers who throw the boxes around the van with impunity.

2022 May. Early budding

2022 July. Leaf strip completed

2022 July. Long view from the opposite hill

2022 July. Canopy all neatly trimmed

2022. Robert in 'canopy work' mode

2022 Auguast. Early evening view

2022 August.Early morning with balloon overhead

2022 Summer. The tasting room awaits

2022 September. Great Pinot going through veraison

2022 Harvest. The team take a break

2022. Crates of gleaming grapes

2022 October. Bird netting in place to deter (look closely!)

2022 December. Frosty vineyard

A Wet and Tricky Season

January 2023
This is the month where the vineyard rests but there is plenty of work to be done in the winery plus lots of planning for the coming season. The most important task this month is to agree with Daniel the dosage for the 2019 brut reserve. The first release of the 2019 will be disgorged this month.

February 2023
Winter pruning commences. A gruelling month of removing last year's canopy, selecting this year's fruiting cane and spur (next year's cane). Lots of tendrils wrapped around the wires need to be cut off carefully. Inspect each trunk for disease and heated wire for damage. We pack the cuttings into Grundon bags and drag them away for burning.

March 2023
Day one. Our trusty Romanians are here seen tying the vines with strict instruction on what to tie to the fruiting wire and what to leave until after the danger of frost passes. I do the tricky ones which might break. Dave Morris and I inspected all the vines before the gang came in and pegged the vines to leave for later. Leaving a vertical 'sacrificial' cane is an insurance both against frost – a second option – and also slows down bud burst throughout the vine.

May 2023
I record, with some surprise, that for the first time I can remember we seemed to have got through a spring without a major incidence of frost. It must be a first. March and April were cold with south westerly winds as opposed to high pressure from the north which tends to bring dry and warm days followed by clear, starry nights……and then havoc at dawn. For the last 3 weeks we have had constant dry north-east winds which can rattle the bones.

I have been bud rubbing the vines the last few weeks and feeling distinctly like 'Toulouse Lautrec' in stature as I shuffle down the rows on my knees rubbing off unwanted buds from the trunk and the crown. The end of May and beginning of June have brought prolonged sunshine with the vines really motoring. What does June hold in store? A flaming month would be very good.

For the first time, as an experiment, we have erected bird netting around the perimeter of each plot in a bid to ward off the attention of the local deer population who like to come in and breakfast, lunch, and dine off the tender young shoots. It seems to be working well apart from the odd muntjac who crawled under the bottom of the netting. They have, since the Pandemic, when culling was reduced to nothing, become a serious problem in the countryside. We need to eat more venison – it is so very healthy.

At the beginning of the month, we could have been spotted applying 'dynamised' horn manure by flicking large droplets with a paint brush from side to side down each row. It sounds a bit wacky I know but heavily diluted manure in water acts to accelerate soil improvement. Our worm life is certainly improved.

June 2023
The highlight of the month was a trip to Offbeat Winery to meet up with Daniel and Nicola (and Jarvis the cocker spaniel), our winemakers, in order to sample the 2022 wines prior to bottling (tirage). These have been vinifying since last October and are now sitting in oak barrels. We were expecting, given the harvest of lovely fruit last year, to taste some exciting wine and we were not disappointed. The 2022 fruit was pressed the same day as harvest, and then underwent first fermentation with the aid only of wild yeasts – no sulphur, no enzymes, and no need for chaptalisation (extra sugar) as is Daniel's style of winemaking.

All eyes, noses, and mouths at the ready as Daniel extracts wine with a wine thief from each barrel for us to taste. The cuvée (the first pressing) and the taille (literally the 'tail', i.e., the second pressing) of the Pinot and Chardonnay wines are held in separate oak barrels. We also have some co-fermented Chardonnay and Pinot in an overflow steel tank; and finally, a separate 228 litre barrel of special Chardonnay which was kept aside to make either a still wine or a blanc de blancs.

A still Chardonnay or a sparkling blanc de blancs? Which to plump for? The fruit was superb. After a little dithering, we decide to produce a small run of blanc de blancs (100% Chardonnay) – something I have wanted to do for a long while. To be able to make this style of wine one needs beautifully ripe Chardonnay grapes. Daniel declared that he was confident overall of making some fine wine. A sense of relief all around. Therefore, we will make a brut, a brut rosé, 300 bottles of blanc de blancs, and later on, a demi-sec. All done and dusted within the hour. Flowering started the second week of June and was largely over in three weeks, thanks to some fine weather. This was followed by excellent fruit set.

July 2023

Another key task – and a long and tedious one – is to work down every row and untangle each vine that has blown into its neighbour. This happened as a result of the wind and the rain we experienced in July. Once upright we insert a biodegradable clip on the wire to keep the shoots apart. If a vine's leaf is partly shaded by another, it follows that the sun cannot reach it directly. This will impair good photosynthesis.

September/October 2023

There were shredded nerves in the run up to harvest this year – *plus ça change*. Life in the vineyard would not be the same without the regular existential crisis. We began to wonder in late September whether the grapes would even reach the required sugar levels before the autumn closed in. Leonie says that I say this each year! We took a bet on the harvest dates…….and I lost my £200.

In late September, we noticed a plague of fruit flies had invaded the Pinot field. It was some years ago when we had a similar outbreak. Vinegar traps were put out to catch them but the damage to some of the bunches had been done. The Spotted Winged Drosophila female has the affrontery to lay eggs under the skin of the berry that then go on to hatch. Red skins are more prone than white ones. To the eye the berry exhibits a brownish colour, is squishy to the touch and has an exit point. This leads to sourness.

As a result, we laboriously spent days on our hands and knees, extracting each damaged berry. The effect on the bunch if not picked out can be ruinous to the crop. We hear from the wine maker that many vineyards are infested. Why? Maybe a perfect storm with the warm and wet conditions experienced over the summer.

If you ever visit the ruins of Tintern Abbey (it is a must), there is a steep vineyard perched on hills above the river Wye. This is one of the vineyards that Dave Morris has taken over from an owner who has decided to retire. I do not blame him, it is hard graft. The vines are well established but they show signs of neglect. We visited this summer. Goodness, what a magnificent view. This could almost be the Welsh equivalent of the famous Hermitage Hill in France, looking down on the Rhône as it snakes its way through the valley bottom. You need to be vigorous and strong in body to tend these vines. Hooves are preferable to shoes.

The dominant characteristics of this season have been low sugars and low acids – a very unusual combination caused perhaps by the heavy rain in July, August and September. While we have had warm days it has often been overcast such that the vine leaves have not had enough direct sunshine which they need to store up sugars.

The yields, as reported in the media, have been high with heavy bunch weights. Time will tell whether the quality is as good as last year. We will not know until next year, once the must has fermented and sat quietly in the winery in oak barrels, whether 2023 will turn out to be a good or an excellent vintage.

Our yield was a good one: 4.8 tonnes across the two fields, with the Pinot producing a little more per acre than the Chardonnay. We found we were very short of oak barrels – this is our preference over steel tanks. We have been struggling to purchase used white wine oak barrels from Burgundy as the French are holding onto them for longer in a bid to save cost. New oak barrels are very expensive, nor are they ideal as they have too much oak influence. The cooperage industry seems to be struggling to meet demand. However, finally, a fellow producer, Hugo Stewart, managed to find a few suitable ones from a good supplier and they were transported in time to take our juice. The reason we ferment in used oak barrels? Good surface contact for softening the acids, for sure. We would always use French barrels from Burgundy over new American oak.

We use mainly 228 litre barriques with a couple of the larger 500 litre demi-muids. This gives the wine maker a wide range of blending choices. Daniel, as a matter of course, will ferment the cuvee and taille of each variety in different barrels.

His style of fermentation is to add no sulphur, no enzymes – this is counter to what happens in conventional winemaking.

Season Report

The start to the season was nigh perfect: no April frosts (a first!) therefore lots of buds; an excellent flowering during a hot June. Then the weather went downhill: an incredibly wet July when it rained every day, mixed weather in August and September with only mid-range temperatures.

The management of the canopy – keeping plenty of air circulating by pruning and thinning – was a headache as we tried to keep mildew at bay. Being an organic grower – we use no synthetic fungicides – there are few preventatives in our armoury. However, it seems that conventional vineyards fared no better this season, even worse in some cases. We can react to heavy rainfall with sprays which battle against the disease forming in the vine leaves – willow bark teas, horsetail teas, yarrow.

The Harvest

This was our 9th harvest, so we know the drill. Check the picking crates, check the number of snippers, buckets, the trailer, the straps. The Pinot harvest took place on 15 October with a great bunch of friends and locals attending at short notice, ensuring that the crop was picked cleanly with next to no disease in the bunches. It needs to be done by careful folk. Green harvesting by Leonie and me before made picking very straightforward. The Chardonnay vendage followed a week later, with an even bigger turnout of helpers, such that we were picked well before lunch. Perfect timing.

There has been much talk in the industry about the introduction of machinery to pick grapes in the light of the problems getting sufficient labour for harvest. Champagne varieties need to be picked delicately by hand— whole bunch with stalk and all. A machine will strip the bunch leaving just berries, and in the process split many of the skins.

Not a problem for making still wine but too brutal for sparkling. Whole bunch pressing ensures a slow juice extraction. The French do not permit picking by machine in Champagne with the result that 27,000 seasonal workers are taken on for the vendange each year. A big cost.

30 October. I am standing in a petrol station in Downton, Salisbury, feeling pretty satisfied having picked up the empty crates from the winery after a very successful operation. The juice is all pressed and moved to barrel having been racked – the solids left in the bottom of the tank and then disposed of.

I am not concentrating. Minutes later it dawned on me that I had put 30 litres of petrol into the tank of my diesel Defender. Two hours later, after numerous calls, it is pumped out by Fuel Fixer and I am on my way home, chastened and £300 poorer.

November 2023

This is the time of year to do nothing, or at least next to nothing, in the vineyard. Except one final biodynamic spray of horn manure plus compost mix. I pick a day when there is no frost on the ground — this is very important. I mix the manure (100gms) into the water and then spend one hour of stirring one way and then the other to create a vortex and thereby 'dynamising' the solution. Petroc Trelawny on Radio 3 keeps me company. I divide the vineyard into sections, then out with a large paint brush and bucket: I walk the rows swishing the brush from side to side and casting large dollops of water onto the grass – the same movement as if one were broadcasting seed with an old-fashioned fiddle.

The vines are shedding their yellow leaves, the atmosphere is damp, the light is dull. A typical November day. Cannot wait to get inside into the warm. Hard to think we were picking grapes just one month ago.

We have been looking high and low for more second hand oak barrels from Burgundy. This is the vessel of choice for fermentation as the internal surface softens the acids and the micro-oxygenation that occurs between through the staves is ideal. There is a shortage as Burgundy wineries are saving money by hanging on to older barrels they would normally sell on after a certain number of years. New oak would be too intrusive for us. The price for a second-hand 228 litre barrels with 3-5 fills has soared in 3 years by 100 percent such that we are now having to pay £500 per barrel. Hugo Stewart of Domaine Hugo – Hugo has great connections in France having lived there or 20 years – secures a batch of high quality and we piggyback on his order. I go down to Offbeat and inspect the new purchases and lovingly glide my hand over the surface. It is like buying a vintage sportscar.

November and December are the months of the year for reflecting on the twists and turns of the season. Could I have done better in managing the canopy against disease; a time for general contemplation and of course planning for the next season. I write up the vineyard diary which records the activities, compares timings of what we did when against those in 2022; then plan next year's winter pruning, and finally check that all the spray records are complete and accurate. Moreover, it is a time for talking leisurely to fellow vignerons about the vagaries of the season. It was without doubt a terrible year for fruit fly devastation. Next year we will be on guard.

The media hyped the 2023 season forecasting an amazing year with record tonnages, fuelled no doubt by the PR departments of the large wine producers. The print journalists forgot to twig that volume does not always equate with quality. The industry was woefully caught on the wrong foot with a lack of sufficient seasonal workers hired and tonnes of grapes went unpicked. Many wineries found they were scrabbling for extra tanks at the last minute; the WineGB forum advertised lots of folk trying to sell excess grapes as they just had no capacity in the winery to press them.

The vines are beginning to shed their yellow leaves, the atmosphere is damp, and the light is dull all day. I cannot wait to get inside close to the warmth of the Aga.

I check in with the winery and all is well. The wine has been 'racked', i.e., transferred to fermentation barrel at which point the solids are left at the bottom of the tank and then discarded. Daniel will use an overflow steel tank with a moveable lid for topping up the barrels during 2024. We place just 500 litres of our wine in a tank with the capacity of 1,100 litres. Otherwise, it is all oak.

The wooden barrels are all carefully marked (last year's markings are scratched off the hoops with a knife) and then lifted into steel racks.

Large producers may not want the outlay and storage for this high percentage of oak barrel fermentation – they might use 5% of oak whereas we use 80% – because we CAN! Small is definitely beautiful.

Interestingly, the biodynamic movement was founded by Rudolf Steiner, the Austrian philosopher and social reformer, in 1924, a full 20 years before the better-known Soil Association was formed. Both movements had similar concerns surrounding the health implications of increasingly intensive farming and its growing use of artificial fertilisers to increase yields. Both groups feared the negative impact of high nitrogen levels on soil quality. How right they were.

I had a fascinating conversation with Tom Petherick, my biodynamic inspector, this summer on the very subject of fertiliser. He told me that before a single shot was fired in WW1, the Germans and British were already squabbling over the vast deposits of guano in South America, so vital was it to the war effort in providing nitrogen, ammonium etc. Towards the end of the war a German scientist by the name of Bosch invented a synthetic fertiliser in the laboratory – the very same name behind the Bosch washing machines and their ilk. I occasionally think of him as I unload the dishwasher.

December 2023

Another month of light duties. As I walk Ludo in the morning, I wander down several vine rows, thinking about how I might tackle pruning a few tricky fellas. Some have unproductive limbs that will need surgery – with due care that the ensuing dieback from the wound does not spread into the healthy centre of the crown. Better small cuts each year rather than one large one.

Next, I look for the potential cane and spurs to be selected next season. Generally, the vines appear in good nick. I subscribe to the gentle pruning method with some exceptions, always mindful of not hindering sap flow whenever I make cuts into the vine's crown.

I have been asked to write an article on the Berkshire vineyard scene for *The Bridge*, the parish magazine in Hungerford, by its editor, Martin Crane. It's a cracking good title – as it works on both pastoral and topographical levels – and it tugs at the memory strings of the residential magazine group, Hill Publishing, I once ran in London many moons ago. We chose similar titles with the definite article a must, such as *The Hill* (covering Notting Hill and Kensington), *The Wick* which became *The Green* (covering Chiswick and Hammersmith), *The Reach* (Chelsea). Shame, I never managed to launch one called *The Bridge* as it is a catchy title and there were plenty of opportunities in a city with countless bridges. Perhaps, as Hackney has now become a des res area, populated with trendy wine bars and smart terraced houses, something entitled *The Carriage* or *The Marsh* would be a hit today. I would not have been seen dead living there 25 years ago. I was always a Hillbilly.

There is no conclusion to this story. We are by no means at the end of the road. As the 2024 season beckons, buds are breaking into life, the secateurs have been re-sharpened, canopy heated wires fired for frost, equipment checked for the coming season. We have just returned from the winery to taste the 2020 and 2021 wines with Daniel and Nicola Ham feeling uplifted by the exciting vintages ahead. It's all systems go.

2023 February. Christopher pruning

2023 March. Gang tying down new canes

2023 March. Snow the day after cane laying!

2023 March. Trevor's sheep on holiday

2023 March. Last year's cuttings

2023 April. Bud burst is magnificent

Robert bends his back to bud rub

2023 August. Leonie 'tucking in'

Ludo the Vineyard Dog

2023 June. Canopy goes nuts

2023 August. Our thermostat is engulfed by foliage

2023 May. Early leaf out

2023 June / July. Flowering

2023 July. Under the vines, long grass and dandelions

2023 August. This is fruit set

2023 October. Chardonnay crates at Offbeat

2023 November. Fuel fixer. What a clot!

2023 November. Vinyard at dusk

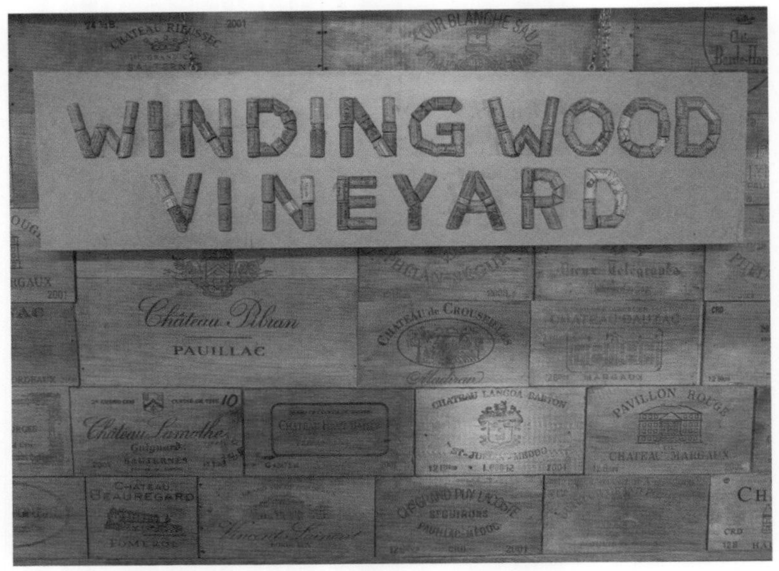

A Lockdown project - the tasting room bar

A view of the farm with a field of rape in the foreground

Appendix 1

Glossary of Terms

Viticulture

Apical Dominance. This occurs when the vine transfers most of its energy to the cane's apex – producing larger bunches at the end to the detriment of the remaining buds. By cutting the apex bud at an angle it should help to stimulate even growth in the cane's lateral shoots. In a cool climate it is almost impossible fully to control.

Biodynamics. See chapter 7. A philosophy of viticulture that not only practices organic methods, avoiding chemical and synthetic treatments, but that also seeks to improve the quality of the soil through the use of preparations and tisanes, as well as through aligning vineyard and cellar work with lunar and cosmic calendars. Many of the great wines of the world are made from biodynamically-grown vines, including those of renowned producers such as Domaine de la Romanée-Conti in Burgundy. The primary organization for biodynamic certification in France is Demeter and this is also the gold standard for England and Wales. Certification for biodynamics in the UK is via The British Biodynamic Association. They also certify for organics as do the Soil Association.

Botrytis. Grape rot, which in terms of wine can come in both desirable and undesirable forms, although in cool climates it is the latter. Botrytis cinerea is only desirable when it takes the form of ‘noble rot’ for making sweet wines, rather than grey rot.

Buds. Typically, we look to allow 9 -10 buds per cane. The first bud to burst is always the most fruitful. If it is destroyed by frost or by other means, then a secondary bud is produced and sometimes a third. These later buds are never as productive and will lag behind primaries in terms of flowering all the way through to ripening, or never get there.

Bunch Selection. One needs a balance between not over-stressing the vine and encouraging fruit. We allow up to 1-2 bunches per shoot but never three, and one bunch on a spur. 2018 was an exception when 3rd bunches ripened perfectly. See *Green harvesting.*

Bud Burst. The buds that form along the cane move from woolly stage to bursting into green and pink buds. This normally occurs towards the middle of April, depending on the weather. Early bud burst can be unwelcome as young buds are vulnerable to clear night frosts. Frost-free years in April are rare. Good site selection is key to avoiding vineyards prone to frost.

Canopy Management. Throughout the growing season the leaf and shoot growth of the vine needs constant managing to keep it thinned and healthy from disease, always allowing plenty of air flow.

Clones. The vine clone type is chosen according to its suitability to take up nutrients in any given soil. At Winding Wood we chose Burgundian clone 95 for our Chardonnay and Burgundian 375 & 385 for our Pinot Noir.

Flowering. The early appearance of the inflorescence and pollination. Tiny flowers appear followed by ‘caps off’, full flowering, then tiny pea sized berries begin to form.

Gentle Pruning. This is a massive subject in itself. The gentle pruning movement advocates making small cuts into the wood and allowing the vine crown to lengthen over the years. Sap flow is a major consideration. Much more difficult to achieve in a cool climate such as in England. Simonit & Sirch are the masters of the gentle pruning technique.

Green Harvest. This is a method whereby bunches which are unlikely to ripen in time are taken off in the run up to harvest. At Winding Wood we do this laboriously in the last months to ensure that by the time it comes to the harvest our pickers can confidently pick all bunches for the crate.

Fruiting Cane. With the guyot pruning system the fruit-bearing cane is replaced each year. This is then tied down to allow its shoots to grow vertically up through the canopy wires.

Laterals. As the vine grows it throws off lateral shoots producing too much canopy. These should be reduced to ensure the energy goes into the fruit.

Leaf Strip. Undertaken in cool climate regions. The leaf around the bunch is removed to maximise sunshine to the fruit. Undertaken normally in August, once the bunch has formed. In a hot climate the grower will go to lengths to protect his fruit from sun burn with leaf shading the bunch.

Nodal Distance. The space between each bud. Uniform distance is ideal. Canes with wide nodal distances will have been affected by extreme temperature shifts in the previous season.

Organic Viticulture. Viticulture practiced without the use of synthetic products such as pesticides, herbicides or fungicides. In northerly latitudes, wet weather poses distinct challenges, particularly when it comes to combatting mildew. A common criticism among conventional circles towards organic viticulture is its continued use of copper. However, there are alternatives to its use. Organic viticulture is steadily gaining a greater following in the UK, especially among the younger generation of growers. Not all organic producers are fully certified. At Winding Wood, we started working organically in 2021.

Potential Alcohol. This is the measure for the natural sugar levels in the grape prior to picking which will convert to actual alcohol via fermentation. Samples of grapes at Winding Wood are taken in the last month in the run up to harvest with sugar levels measured in the berry juice using a refractometer. Either Oechsle or Brix scales are used to determine the reading for potential alcohol.

Pruning System. There are various systems. Most common for sparkling wines are single or double guyot where a new cane is bent in one direction or, in the case of a double guyot, two canes are laid in separate directions. The cane should come from first year wood to ensure fruitfulness. A lower spur is chosen which will become the cane the year after. In Champagne they mainly shoot prune – cordon royat – for Pinot from old wood allowing the body of the crown to extend forward in a permanent cordon. For Chardonnay, an intricate Chablis method of pruning with 3-5 canes is common.

Rootstock. The vine is grafted onto a rootstock which post-Phylloxera will be of American origin. The graft is then protected by wax. Winding Wood used a S04 rootstock throughout, common in the UK. Phylloxera vastatrix is a vine louse that is thought to have originated in eastern North America, but then spread around the world, beginning in the middle of the nineteenth century. The only major regions to avoid this louse were the Canary Islands, Northern Chile and the island of Santorini in Greece. It attacks the vine, destroying its root system and depriving the vine of its ability to intake nutrients and water.

Spur. A spur is a short cane cut to approx. 1-2 inches with two buds showing, ideally sitting behind and below this year's fruiting cane to ensure the sap reaches the spur first. This will become next year's fruiting cane. Two spurs are commonly created, one low down on the front of the trunk and one low down behind

Terroir. The French concept of terroir refers to the place itself and the influence of factors such as soil, microclimate, region climate, plus anything else that could possibly have an effect on the plant in that location. Two neighbouring vineyards, on similar soils, growing the same varieties of grapes can produce different tasting wines. However, what happens in the winery cannot be underestimated in producing unique characteristics in any wine.

Tendrils. The vine produces tendrils which wrap themselves around the wires. A vine, after all, is a woodland plant which wants to climb towards the light.

Trellis System. Each row will have an anchor post at each end, set at an angle and anchored to take the tension and then intermediate posts installed every 6 vines or so. Posts can be wood but more commonly in UK they are galvanised steel with lugs up the sides (to hold the wire) as they last longer in damp conditions. A fruiting wire is run through the trellis at 1 metre above ground with two to three canopy wires above. These can be moved up or down depending on the growth of the canopy. Our frost prevention heated wire is attached to the fruiting wire. The fruiting wire will be thicker in girth to that of the canopy wires.

Tucking In. Procedure for tucking in any vine shoots which grow out horizontally from the main canopy. This ensures that they are not broken when the tractor travels down the rows. From time to time, vignerons will walk down each row 'lifting' the wires away from the posts, thereby snapping the vines' tendrils and collecting all the freed-up canopy neatly within the wires, and then hook the wires back onto the lugs. A job best undertaken by two working either side of the row.

Vendage/Vendange. Harvest. This can take a matter of days or weeks depending on the size of the vineyard. Early ripening varieties include: Bacchus, Reichensteiner, Madeleine Angevine, Rondo, Seyval Blanc, later ripening: classic champagne varieties like Chardonnay, Pinot Noir, Pinot Meunier. Winding Wood tend to harvest our Pinot Noir at least a week before the Chardonnay. The general guide for determining the likely harvest date is 100 days from completion of flowering. Often vineyards undertake a 'golden harvest' taking the ripest bunches first. Determining the best day to harvest is planned with several factors: the weather forecast, grape ripeness, and the winery's schedule of press availability. At peak time some wineries will be pressing around the clock. Sparkling producers pick by hand to ensure the grapes do not get damaged. Champagne employs over 27,000 workers for the harvest. Grapes destined for making still wines are increasingly picked by machine.

Veraison. The stage at which the grape bunch changes colour and consistency and then begins to ripen. Pinot Noir will go from green to deep purple when ripe, Chardonnay from bright green to golden yellow.

Vine Rows. Distance between each row: the standard is 2.4m, whereas at Winding Wood we opted for row widths of 1.8m. Our reasoning was to produce more potential bottles for the acreage. In Champagne there is barely room to walk down each row, as they are mostly 1m apart. The width of row determines the size and power of the machinery employed.

Vine Clips. Biodegradable clips which attach to the wires to hold the vine shoots vertical. Otherwise, they can entangle in the wind and rain. This mangle can produce poor photosynthesis in the leaf.

Winter Pruning. Annual operation of cutting off last year's cane and canopy from the wires, then choosing what will be this year's cane and spur (next year's cane hopefully). Cuttings can be cut and left on the ground in the rows and then rotovated into the grass; or, as we do, for reasons of disease prevention, take remove, chip and compost. Normally undertaken from end of January, through February, and into March either in one or several stages. It is a laborious manual job as the vines will have attached themselves by tendrils to the wires so need cutting cut with secateurs.

Winery

Most vineyards in the UK do not have the on-site facilities to make wine. The general guide is that a minimum of 40 tonnes of grapes annually are needed to make a winery viable. Small and medium-sized vineyards arrange to take their grapes to contract wineries to be pressed, fermented, and bottled. Where possible wineries should be located close to the vineyard so the grape crates can be delivered on the day of picking. Recently we have seen, with a rise in demand, the establishment of more dedicated contract wineries.

Acids. Two main acids found in grapes: tartaric and malic – plus small amounts of citric and a few other acids. Malolactic fermentation is normally undertaken – sometimes it occurs naturally – to convert the malic into lactic acid, thus reducing the overall harshness. After all, it is the taut balance between acid and sweetness that makes sparkling wine and champagne so special. Too much acid gives a bitter taste or after kick, whereas the correct level offers freshness – a characteristic of English bubbles.

Assemblage. The blending of wines of different grape varieties for a particular sparkling or still wine. We use this method to make our brut rosé as the wine maker can obtain the desired taste and colour more accurately than through the saignée method (see below).

Autolysis. The breaking down of yeast cells through an enzymatic process produces a chemical reaction. This occurs when a wine is left in contact with the lees over a long period of time. Autolysis is key in making champagne or sparkling wine, as it imparts particular types of flavour – those of brioche, nut, biscuit – complexity and textural finesse that can be achieved in no other way; and this is one of the main reasons for champagne's long aging on the lees in bottle after the second fermentation. See *Lees Ageing.*

Barrique. The most common size of wooden barrel, almost always made of oak. In Champagne, barriques are historically 205 litres in size, although many producers use old 228-litre barriques purchased from Burgundy or 225-litre barriques from Bordeaux. Larger barrels are generally referred to as demi-muids (500-600 litre), while large casks are called foudres. Winding Wood Vineyard use a mix of 228 and 500 litres barrels.

Base Wine. A still wine, or vin clair, usually fermented to roughly 10 to 11 degrees of alcohol, used as a component in the blending of sparkling wine. Champagne is generally blended from many different base wines. The larger English producers now do the same. Corinne Seeley at Exton Park makes exquisite NV wines from many years of blends.

Bottle. The glass for champagne and sparkling wines must be very strong to withstand the enormous pressure of the contents up to 7 bar of pressure. This is equivalent to the pressure in the tyre of a double decker bus.

Bâtonnage. The stirring of the lees, generally performed with wines ageing in barrel, although it can also be done in tank. The lees are stirred to put them in suspension, giving greater richness to the wine. This practice is common in Burgundy for still wines. Many UK makers of white wines will follow this practice. Offbeat Wines certainly do.

Capsule, or Crown Cap. A metal capsule, as found on beer bottles, which is commonly used today to seal a bottle during the second fermentation and ageing on the lees. At disgorging this is replaced with a cork and wire hood to hold the cork on securely under huge pressure.

Carbon Dioxide. The build-up of CO2 in a winery can be very dangerous to health. The cleaning out of the insides of the tank can be hazardous. At the beginning of each day the winery doors should be left open to allow the gas to escape.

Cépage. Grape variety. The classic varieties in champagne are Chardonnay, Pinot Noir and Pinot Meunier in varying percentages according to house style. England and Wales have largely adopted these varieties to make traditional method wines.

Chaptalisation. The adding of sugar to grape must in order to raise its degree of potential alcohol. The process is named after Jean-Antoine Chaptal, a chemist, who introduced this practice during the time of Napoleon.

Cold-Stabilisation. The controlled chilling of a base wine before bottling in order to cause the formation of tartrate crystals so that they do not occur later in the bottle. This is widely practiced although some wine producers eschew it as they think it an unnecessary manipulation of the wine. Considered essential for sparkling wine.

Cork. The most common material used to close a bottle, and mandatory in champagne for finished wines. Champagne corks are constructed differently from regular corks used for still wines.

Corked. An all-too-common wine flaw due to the presence of TCA (2,4,6-trichloroanisole) in the cork, which imparts a musty, unpleasant aroma. We use a composite Diam cork.

Cuvée. In the world of wine, the word 'cuvée' generally refers to a batch or blend of wines. In Champagne it also has another, very specific meaning: during pressing, the cuvée is the first 2,050 litres of juice from a 4-tonne press, which represents the finest portion of the pressing.

Débourbage. A settling of the juice after pressing to remove solid particles such as skins and pips before fermentation.

Disgorgement. The process of removing the yeast sediments after fermentation and ageing in bottle. The sediment is collected in the neck of the bottle through riddling, whether manually or via the use of a gyropalette. Typically, the neck of the bottle is then frozen to collect the sediment into a solid mass, and then this lozenge is ejected when the capsule is removed. Some growers, however, still disgorge bottles by hand without using the freezing technique. It is called 'à la volée'.

Dosage. See Appendix 4.

Fermentation. In wine, the conversion of sugar into alcohol and

CO_2 by yeast. Sparkling wines undergo two fermentations: the first in tank or barrel which creates a light, still wine; and the second after bottling at which stage a small amount of sugar and yeast is added to ensure a good fermentation. During this second fermentation the carbonic gas, responsible for providing the bubbles, is trapped in the bottle.

Fermentation Vessels. Steel tanks, concrete tanks, oak barrels, carboys, amphora (clay pots). There was an innovative wine maker in Mallorca at Ca'n Vidalet who fermented in sealed amphora under sea water. Flavour full of brine not surprisingly! At Winding Wood we use primarily oak barrels for fermentation. These are by convention Burgundy barrels which have been used for '5 fills' of white wine. We choose oak as it provides ideal micro-oxygenation plus contact for softening the acids – rather than imparting 'oakiness' which we definitely do not want. Sometimes we leave the wine in barrel for longer if we feel it will improve. Every barrel is different which is always a challenge and only tasting can tell what appropriate action to take. Blending is a great skill.

Filtration. A process used to remove suspended particles in a wine. Most wines are filtered before bottling, not only in Champagne but elsewhere in the world. However, some producers choose not to filter, as with Offbeat, as they believe that it removes character from the wine.

Fining. A clarification of wine by the addition of a physical agent, such as bentonite or egg whites, that removes solid matter. Offbeat avoid this process.

Foil. The coloured aluminium foil which covers the wire hood. Now we have left the EU, it is no longer mandatory to use a foil.

Gravity-fed Winery. This approach is gentler with less intervention than some conventional methods. It is important to keep to a minimum, or eliminate completely, the pumping of wine around the winery, to reduce the introduction of oxygen into the wine and, so by, preserve its purity. Gravity-fed wineries are increasingly popular but very expensive to construct. They are built on various floors to ensure the wine moves through each stage by gravitation only, from the pressing itself on the top floor through to the bottling on the ground floor.

Gyropalette. A mechanical device used in place of riddling in a pupitre invented by Madam Clicquot (of Veuve Clicquot fame), to collect the yeast sediments in the neck of the bottle in preparation for disgorgement. Gyropalettes have made this process much faster, with no resulting loss in quality. Almost all producers, large and small, use gyropalettes today, although some still riddle bottles by hand, either to preserve tradition or to accommodate oddly - shaped bottles that don't stack neatly in the cage.

Labels. Typically, there are three labels, although sometimes two, on a bottle of sparkling wine: front, back and neck label. The neck label gives a bit of extra flourish to the bottle's appearance but it also has the function of covering the bottom of the foil.

Lees. The dead yeast cells from the second fermentation which are removed from the bottle at disgorgement.

Lees Ageing. See autolysis. Lees ageing in sparkling wine generally refers to the ageing of the wine in bottle, where the yeast cells left over from the second fermentation are trapped inside the bottle until the bottle is disgorged. This period of ageing on the lees is fundamental to the creation of champagne and sparkling wine's unique character. By law in France, champagne must remain on lees for 12 months and in bottle for 15.

Liqueur d'Expédition. The blend of sugar and wine added to champagne or sparkling wine at dosage, just after disgorgement. Either cane or beet sugar is normally used, and the wine can be young or old, depending on the producer's preference. Winding Wood always use wine from the same vintage.

Liqueur de Tirage. The solution added at bottling to induce the second fermentation, composed of wine, yeast and sugar. As a general rule, four grams of sugar per litre of wine will create enough CO2 to produce one atmosphere of pressure (1 bar); the standard measurement in champagne is 24 grams of sugar, which produces roughly six atmospheres or 6 bar pressure, although a little may be lost at disgorgement.

Malolactic Fermentation. Not strictly speaking a fermentation, but rather a conversion of sharp, malic acidity to softer, creamier lactic acidity. It's widely practiced in the UK to balance the high acidity levels although some producers in Champagne choose to block it in order to retain the firm, lively structure of the malic acidity.

MCR. (Moût concentré et rectifié). Concentrated and rectified grape must, preferred by many for the dosage, instead of the traditional liqueur d'expédition.

Méthode Champenoise. The old name for the traditional method of making champagne, banned by the European Union in 1985 as a concession to protecting the Champagne appellation. Now called Method Traditionelle.

Muselet, and Plaque de Muselet. The wire cage that holds a champagne cork in place, while the plaque de muselet is the metal disc affixed to the top of the cork, that usually has some sort of design unique to the producer. They are very collectible. We have a good collection in the Winding Wood tasting room hung from a sheet of fruit netting.

Non-Vintage or NV. A term used to refer to wines blended from multiple years. Non-vintage champagnes and sparkling wines can be made at many different levels of quality: a non-vintage brut is generally a producer's basic, entry-level champagne. At Winding Wood, to date, we have only produced vintages to date, ie grapes from that year's harvest.

PDO & PGI. In England and Wales, acronyms for Protected Designation of Origin and Protected Geographical Indication. This identifies a product that originates in a specific place, region or country, the quality or characteristics of which are essentially or exclusively due to a particular geographical environment with its inherent natural factors. Sussex controversially has declared its own appellation. Other regions employ similar acronyms.

Post-disgorgement Ageing. Or ageing 'on cork'. Champagne and sparkling wine is aged for much longer in the cellar than most still wines. Top-quality producers try to hold their champagnes after disgorgement for at least six months before release, as the wines need time to recover from the physical shock of the act, but champagne almost always benefits from additional ageing beyond this. Even basic, non-vintage bruts can often improve with another year of ageing, while vintage champagnes can sometimes continue to evolve for decades. It should be noted, too, that ageing before disgorgement and ageing after disgorgement achieve different results, as the wines exist in different environments: before disgorgement, the wine is in a largely anaerobic environment, as the lees are natural antioxidants.

Pupitre. Wooden 'A' frame rack, invented by Madame Clicquot (of Veuve Clicquot) for riddling bottles. In Champagne, this used to be a dedicated job with the operator able to twist 5,000 bottles a day. Probably not good for repetitive strain injury. The bottle capacity for a pupitre is 120 bottles. We have one for show in our tasting room.

Racking. Moving the must from one vessel to another in order to leave behind the solids at the bottom on the tank.

Reserve Wine. Older wines that are blended with the current harvest to make a non-vintage sparkling wine. Smaller growers usually keep reserve wines from the last one or two vintages, while some négociants have large stocks of reserves dating back for decades. Louis Roederer, Bollinger and Krug are among those houses famous for maintaining a vast collection of reserve wines.

Saignée. Literally, 'bleeding'. In sparkling wine and champagne, a process of making rosé champagne in which colour is derived from maceration on the skins rather than with blending of red wine. The saignée method tends to produce rosés that are darker in colour and more pungent in aroma.

Sparkling Wine. Any wines of an effervescent nature. These include champagne, English sparkling wine, crémant, cava, and prosecco. The word 'Champagne' refers to a geographically delimited appellation, just as Bordeaux refers to wines from a region in southwest France and Napa Valley refers to a region in northern California.

Stillage. Metal-framed cages for holding approx. 500 bottles tightly on their sides for lees ageing. These are then stored in high stacks in the winery and removed with care by a forklift.

Sulphur Dioxide. Used to preserve wine and keep it from oxidising. Organic and biodynamic wines use very little or none. It is a requirement of the Foods Standards Agency for producers to display on the back label if the wine contains sulphur dioxide ('sulphites') – but not on a packet of crisps or biscuits which contain far more!

Taille. Refers to the second pressing of the grapes, composed of 500 litres of juice directly after the initial pressing of 2,050 litres called the cuvée. The taille is generally considered to be inferior to the cuvée, but some producers choose to use a small quantity of taille for its fruitiness and its lower acidity.

Tartrate Crystals. A precipitation of tartaric acid that occurs at low temperatures. Tartrate crystals look similar to sugar crystals and are completely harmless, with no detrimental effect to either the wine or the wine drinker. However, in an effort to spare the consumer the sight of such things, most sparkling wines undergo a process of cold stabilization, which forces the crystals to precipitate before bottling.

T/A. Titratable acid. Measure for the total acid (largely tartaric) contained in the fruit prior to pressing.

Tirage. The act of bottling. At Winding Wood, we leave the bottling until August the following year after harvest – longer than most – and then clean the oak barrels in time for the autumn harvest. However, many of the large wineries in the UK use French mobile bottling companies which come over in the late spring each year bringing their equipment with them. Some now keep a machine in the UK which tells you something.

Vin Clair. A still wine resulting from the first fermentation in tank or barrel, before it is bottled and transformed into sparkling wine.

Vintage. The year of harvest. These are champagne or sparkling wines made from the harvest of a single year (technically 85%), instead of being blended with wines from multiple years – as with most champagnes. Vintage champagnes tend to undergo longer ageing and are only declared in great years.

Wine Press. There are three main types of grape press: traditional hydraulic wooden basket presses like the Coquard (Offbeat own one) which is the Rolls Royce of presses; bladder presses of all sizes which are the most common; and small basket presses for low volumes of grapes (under 200 kilos) similar to apple presses in look that are manual. The small basket press is perfect for small batches of grapes but one needs a good set of muscles to operate. Of course, the expensive wine press, like the farm combine harvester, will sit idle for 11 months of the year.

Yeasts. Natural or cultured yeasts required to aid both fermentations.

Styles of Champagne and Sparkling Wine

Blanc de Blancs. A wine made exclusively from white grapes, which almost always means that it is 100% Chardonnay. However, champagne can, but it is rare, be made from other white grapes, such as Pinot Blanc, Arbanne or Petit Meslier. Peter Hall, of Breaky Bottom in Sussex makes a blanc de blancs from Seyval Blanc which is delicious.

Blanc de Noirs. A sparkling white wine, sometimes with a pinkish hue, made exclusively from red grapes. In Champagne, it can be 100% Pinot Noir, 100% Pinot Meunier, or a blend of the two.

Brut. The most common style containing 0-12 grams of dosage/litre. Usually a blend of 2-3 grapes: Chardonnay, Pinot Noir and Pinot Meunier in various ratios. We are Chardonnay-dominant having planted 60% Chardonnay and 40% Pinot Noir. Most champagnes, interestingly, are red-fruit dominant.

Brut Rosé. A pink version of champagne or sparkling wine, more often than not, made by the blending of a little red wine with normally vinified base white wine. Rosé sparkling can either be made using the 'assemblage' method that involves blending wines; or by the 'saignée' method which involves extracting precise colour from the juice at the time of pressing.

Crémant. A style of bottle-fermented sparkling wine made at a slightly lower pressure than traditional champagne, usually 3.5 to 4 atmospheres (bar) instead of the standard six. And usually less time on lees. This term is now used for sparkling wines in other parts of France. The areas of France best known for their crémants include Bourgogne (Burgundy), Limoux, the Loire, Jura, and Savoie.

Demi-Sec. Made in the same way as a brut, but with a much higher dosage added at disgorgement – around 35/38 grams per litre. It is an unusual wine but ideal for accompanying fruit-based puddings. Winding Wood started making a demi-sec at the suggestion of our winemaker at the time, Emma Rice, and it has been well received. We only make between 100 and 150 bottles per year. I call it our ‘nightclub wine’ – to be sipped after supper into the wee hours.

Col Fondo. These wines are more natural, much less sweet than your average prosecco. Unlike champagne or sparkling wine, there is zero disgorgement (removal of the frozen sediment/yeasts), making the wine cloudy and funkier looking. Loved by organic wine drinkers.

Pét Nat. Abbreviation for pétillant naturel. Increasingly popular with young organic wine makers and drinkers. This is a single fermentation in a bottle with a crown cap which produces a gentle fizz. Much cheaper to make as it is released young.

Appendix 2
Grape Stages - From Bud to Grape (1-47)

Month		Vine Stages 1-47	Land care
December Leaves have fallen from the vines leaving bare canes. The energy of the plants has moved from above to below ground and the soil biome continues to work throughout the winter.		47-1. Dormancy winter bud	None
January A quiet time, no growth above ground in the vineyard, too wet to work on the soil, a time for the farmer to rest, reflect, recuperate and plan ahead.		1. Dormancy winter bud	Take soil samples every other year.

Month		Vine Stages 1-47	Land care
February Another quiet month and yet a time of hope and beginning to look forward and see the first signs of life. Begin to winter prune on dry days.		1. Dormancy	Pruning Drop lower canopy wire to aid tying canes at a later stage.
March The beginning of spring, the first flush of flowers, the sap is beginning to rise and the early tree buds are bursting.		1. Dormancy winter bud. Leading to woolly bud stage as the bud begins to swell.	Pruning/sheep grazing. Finish pruning and vine tying. Crown cuts when sap rising. Check functionality of heated wires.

Month		Vine Stages 1-47	Land care
April Mid spring, meadows of spring flowers, leaves are coming out, first butterflies, birds are nesting, a time of increasing energy.		2-6. Bud burst. The most vulnerable stage with cold, clear nights giving way to early morning frosts.	Red alert for overnight frost.
May The height of spring moving towards summer; the lime- green of new leaves everywhere, vibrant and new, full of upward life energy.		7-18. Early shoot and leaf growth.	Bud rubbing after last frost. Leave crown buds to inhouse team as delicate work.

Month		Vine Stages 1-47	Land care
June The midpoint of the year, all the buds have burst and the vine shoots and leaves are growing rapidly in the longer hours of sun. A time of strong upward growth.		7-19. Early shoot and leaf growth and beginning of flowering.	Leaf stripping depending on flowering stage. Strim under vine trunk. Canopy reduction.
July The month of flowering, moving towards the height of summer, flowers are transforming to become the first fruit, while grain is ripening.		19-27. When flowering is complete this stage is followed by the beginnings of fruit set.	Leaf stripping, strim, tuck in vigorous shoots. Reduce 3rd bunches. Canopy reduction Mow alleys. Check for fruit fly (SWD).

Month		Vine Stages 1-47	Land care
August The month of fruit set, the first chance to see the potential harvest, moving towards the end of summer, the vine begins to focus its energy into the fruit.		27-35. Berry formation, with swelling and ripening of early varieties.	Tucking in, mow alleys. Check for fruit fly (SWD).
September The first month of ripening, the first potential harvest, a time of gratitude and celebration.		34-38. Veraison. This is the softening and ripening of berries. In the case of our Pinot, the grapes begin to turn to red and then black.	Tucking in, bird netting on. Start the chore of green harvesting.

Month		Vine Stages 1-47	Land care
October The 2nd month of ripening and harvest. The beginning of leaf fall for early varieties, later in the month.		37-39. Ripening of berries. Stage 38 = time to harvest all ripe bunches.	Bird netting, green harvest, then harvest. Clean picking crates.
November Leaf fall for mid/late varieties. Temperatures usually dropping significantly with rainfall increasing.		41-47. After harvest, beginning through to the end of leaf fall.	

Appendix 3 - Stages of Wine Production

Winding Wood vintages in production at Offbeat Wines as at June 2024

Style of wine	**Pressed & fermented**	**Bottled for 2nd fermentation**	**Months of lees ageing in bottle**	**Riddled, disgorged & dosage**	**Released after 'cork' ageing**
Brut Reserve 2018	Oct 2018	April 2019	38 months plus	June 2021 (1st release) June 2023 (final)	1st release Dec 21
Brut Reserve 2019	Oct 2019	March 2020	50 months plus	Dec 2022	Mar 2023 (club members only)
Brut Reserve 2020	Oct 2020	Aug 2021	32 months		
Brut Reserve 2021	Oct 2021	Aug 2022	20 months		
Brut Reserve 2022	Oct 2022	Aug 2023	11 months		

Style of wine	Pressed & fermented	Bottled for 2nd fermentation	Months of lees ageing in bottle	Riddled, disgorged & dosage	Released after 'cork' ageing
Brut Rosé 2020	Oct 2020	Aug 2021	18 months plus	Feb 2023 (1st release)	June 2023
Brut Rosé 2022	Oct 2022	Aug 2023	11 months		
Blanc de Blancs 2022	Oct 2022	Aug 2023	11 months		
2023 Vintage	In barrel, 1st fermentation	Aug 2024 blended and bottled to produce Brut Rosé and Blanc de Noirs			

List of wine medals won in competition

WineGB Awards 2024 Gold	**Brut Rosé 2020** WineGB Awards 2024 – GOLD and SPARKLING ROSE TROPHY	IEWA Independent English Wine Awards 2022 Silver	**Demi-Sec 2018** Independent English Wine Awards 2022 – SILVER
IEWA Independent English Wine Awards 2024 Gold	Independent English Wine Awards 2024 – GOLD & TOP SPARKLING TROPHY	IEWA Independent English Wine Awards 2021 Silver	**Brut 2017** Independent English Wine Awards 2021 – SILVER
IEWA Independent English Wine Awards 2022 Gold	**Brut Rosé 2018** Independent English Wine Awards 2022 – GOLD	WineGB Awards 2020 Gold	**Brut 2016** WineGB Awards 2020 – GOLD
WineGB Awards 2022 Silver	WineGB Awards 2022 – SILVER	The IEWA Independent English Wine Awards 2020 Silver Medal	Independent English Wine Awards 2020 – SILVER
WineGB Awards 2024 Gold	**Brut 2018** WineGB Awards 2024 – GOLD	2020 Decanter World Wine Awards Bronze	Decanter Awards 2020 – BRONZE
IEWA Independent English Wine Awards 2022 Silver	Independent English Wine Awards 2022 – SILVER	Thames & Chilterns Vineyards Association Silver Wine of the Year Medalist	Thames & Chilterns Vineyard Association 2019 - SILVER

Appendix Four

Dosage

Once disgorging has taken place and the yeast lozenge expelled, the wine will be clear. The bottle will also need topping up as it will have lost liquid given the pressure. Adding dosage at this stage is normal practice. It is a good additive, the cherry on the cake, and it allows the winemaker to fine tune the finished wine.

Dosage is a sweet mix of sugar and wine called liqueur d'expédition which makes the wine more balanced and complex. Either concentrated and rectified grape must or sugar cane is used by the winemaker. Although dosage is a time-honoured process – champagne was originally a sweet drink – over the years the trend has been to use less sugar giving it a dry finish. Indeed, zero dosage is popular with organic and biodynamic wines.

The normal range for a brut is between 6 to 12 grams of sugar per litre. One would be amazed how much difference 1 gram of sugar makes to the taste. When we do blind dosage trials with the winemaker this is regularly the case.

Justin Howard Sneyd (MW) of Domaine of The Bee and Heart of Gold, writes interestingly on the subject of dosage. He thinks that brut zero can be' joyless' and muted like 'singing under a duvet'. I think somewhere around 3gm/l can be a perfect balance depending on the year. Interestingly, in Spain, cava winemakers sometimes use sherry. In Champagne, it is not uncommon to add cognac to the sugar.

1. Brut Nature or Non-Dosé. Contains zero dosage or less than 3 grams per litre. Bone dry

2. Extra Brut. Must contain no more than six grams of sugar per litre. Champagnes between 0-6 grams per litre of dosage also qualify for the Brut designation, although most producers will label them as Extra Brut rather than Brut. Bone dry

3. Brut. The most common style containing anywhere between 0-12 grams per litre of dosage. Note that wines of between 0-6 grams per litre can be called either Brut or Extra Brut. Dry

4. Extra Sec or Extra Dry. A dosage level of between 12 and 20 grams of sugar per litre. Sometimes called 'Fruity'. Not for me.

5. Dry. A dosage level of 17- 32 grams per litre. Off-dry category but 'dry' does not come into it and rarely seen except with very unripe fruit perhaps.

6. Demi-Sec. A relatively sweet style of champagne or sparkling wine, containing a dosage of between 33 and 50 grams of sugar per litre. Winding Wood produce a demi-sec with 38 grams per litre. We call it our 'nightclub drink'. Sweet but with plenty of acidity to be had throughout the evening after a meal or delicious eaten with a fruit pudding. Utter heaven with a piece of chocolate brownie.

7. Doux. This is the sweetest of the official categories of champagne, used to refer to wines with a dosage of over 50 grams per litre While this was the most common style of champagne in the 18th and 19th centuries, it's virtually non-existent today.

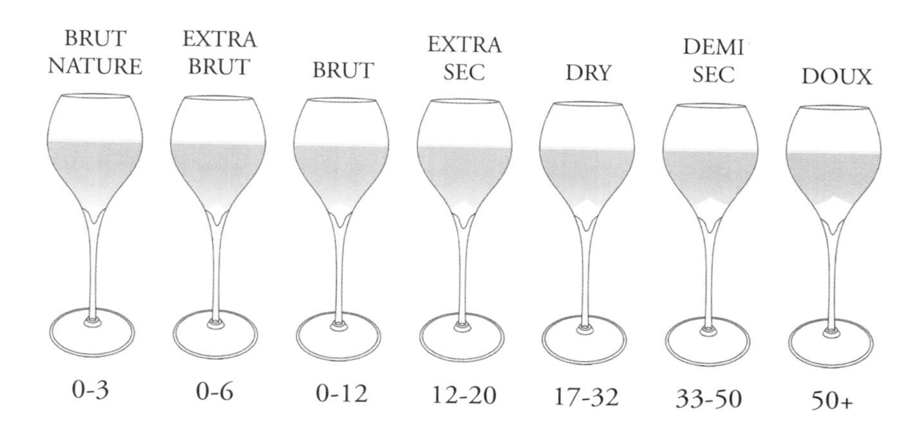
SWEETNESS LEVELS IN SPARKLING WINE
BRUT NATURE
EXTRA BRUT
BRUT
EXTRA SEC
DRY
DEMI SEC
DOUX
0-3
0-6
0-12
12-20
17-32
33-50
50+

Appendix 5

Biodynamic Preparations

There is, more often than not, a giggle from visitors during our wine tours when I describe the medicinal plants from which we make teas for spraying on the vines as protection against mildew. I mention plants such as willow, horsetail and yarrow. There then follows a chuckle when I throw in how we use small quantities of cow manure that have been pressed into horns and buried over winter. Next, barely concealed amazement when I say that we mix this ingredient with water for an hour (while chanting) and then broadcast over the vineyard by flicking onto the grass with a large paint brush – the same action as a back-handed tennis stroke. Then follows outright guffaws when I describe how we carry out various tasks according to the phases of the moon. Mon Dieu! I wiggle my ponytail and move on.

Rudolf Steiner, the father of biodynamic agriculture (and more besides) described in his lectures in 1924 nine biodynamic preparations from natural substances: cow manure, quartz and seven natural substances – all easily available for most farmers.

They are:

- Horn Manure (BD 500)
- Horn Silica (BD 501)
- Yarrow (BD 502)
- Chamomile (BD 503)
- Stinging Nettle (BD 504)

- Oak Bark (BD 505)
- Dandelion (BD 506)
- Valerian (BD 507)
- Horsetail (BD 508)

These preparations are transformed from their natural base state in various ways: some are sheathed in animal organs; some are buried, hung in the air, or soaked in water; all are made at specific times of the year; all are transformed through specific periods of the year. The liquids are 'dynamised', which is to say stirred rhythmically with a paddle in a vessel for one hour. There are copper kettle dynamisers with motorised blades available on the market, at a fancy price, which will do the stirring for you but that rather defeats the purpose of taking an hour out of a busy day to do the mixing in quiet contemplation.

These preparations 503-508 come from the plant kingdom, albeit some are first fermented in animal material. They are referred to by name of the plant or by the numbers given by Steiner. These are used to make composts of organic matter – composting in rows or heaps and covered with straw, hay, and earth before specific preparations are introduced. Biodynamic preparations, be they for spraying via teas, decoctions, extracts or just introduced into composts, are a basic requirement for promoting good plant health. These plant extracts not only help prevent several pests and crop diseases but can act as a curative. They are readily available and often found locally – a key to their usage.

In the northern hemisphere, horns are filled with manure and silica from March to April. These then remain in the earth for six months from autumn to spring equinox (psto provide winter energy) until September/October when they are dug up.

BD 500 & 500p –Horn Manure. Field Spray
Sourced from good cow manure, scooped into a horn and buried over winter. Helps to build soil structure, stimulates microbial activity, regulates the soil acidity, stimulates growth of the root system. It is diluted in water and stirred energetically (dynamised) for exactly one hour. Best applied on a day that is neither too hot nor too windy. 100g of horn manure in 35 litres of water per hectare of land.

BD 501 – Horn Silica. Field Spray

Made from rock crystal and crushed into a powder. Essential preparation for all biodynamic agriculture. Complimentary and in polarity to horn manure. Brings a crystalline quality to plants, mitigates disease, and reinforces the effects of sunlight on the vine's canopy. Best time to spray is just after sunrise and preferably before 8am. 2-4 grams in 25-35 litres per hectare. Stirred energitically with a paddle in the same way as horn manure. Best applied with a knapsack and extended lances.

BD 502 – Yarrow Preparation *(Achillea millefolium)*. Compost prep

Plays an important part in the processes connected with sulphur and potassium, and a secondary role regulating selenium and silica. Plant can be found easily enough in lawns or hedgerows. Tellingly, even in wet conditions, yarrow always looks fresh. Tea made from yarrow helps to reduce the amount of spraying needed with sulphur. It is said that yarrow sprays help to make finished wine less susceptible to oxidisation.

BD 503 – Camomile Preparation *(Camomile recutita)*. Compost prep

Chamomile is often cultivated as a medicinal plant. It is connected to the metabolism of calcium and regulates the nitrogen process through reduced loss of ammonia. It has a secondary role in regulating potassium, boron and manganese. The flowers can be harvested as soon as the petals are fully extended, and then dried.

BD 504 – Nettle *(Urtica dioica)*. Compost prep

Stinging nettle tea is a regulator of vegetative growth. It regulates both nitrogen and iron and in a secondary role, potassium, sulphur, calcium, magnesium and manganese. Moderate on its own as a preventative against mildew, it is best incorporated in with other plant extracts. Know in viticulture for controlling spider mites. It is a good additive to compost encouraging the breakdown of organic matter into humus. The ideal time to harvest nettle is at the beginning of flowering stage. A few handfuls are placed in cold water and simmered. As a biodynamic tea, nettle is used frequently in viticulture in a mix with sulphur and copper.

BD 505 – Oak Bark *(Quercus robur)*. Compost prep
Oak bark gently mitigates fungal disease in plants and has a special relationship to calcium. We tend to use willow bark as an alternative (see below). Only use outer bark from standing or newly felled trees.

BD 506 – Dandelion *(Taraxacum officinalis)*. Compost prep
This helps to regulate silicic acid and hydrogen, influences the processes of potassium, limestone, and nitrogen. We pick the heads and dry them, from which we make a tea.

BD 507 – Valerian *(Valerian officinalis)*. Compost prep
Grows between June and July, usually along streams, damp meadows and on the edge of forests. Only the flowers of the plant are used in viticulture and agriculture. Valerian, using the whole plant, was once known as a remedy for heart conditions, stress, and insomnia in humans. The flowers are pressed extracting a dark juice with an animal smell. Very useful in the spring especially if the plants are stressed due to variations in the climactic conditions.

BD 508 – Horsetail *(Equisetum arvense)*. Field spray
The stems and leaves are dried and then made into a decoction. Simmer with water from a few minutes to an hour over a low heat with covered pan. It is used as a preventative measure against mildew, rust, and other fungal diseases. It is the master of damp conditions hence its effectiveness against fungal attack. It is rich is silica which is an aid in strengthening plant tissues. Best used in times of heat and humidity, especially close to a full moon and the moon's perigee. It must be used with care in the growing season as it is a powerful and can dry out plants.

Willow Bark Tea *(Salix)*. Field spray
We tend to use willow over oak bark as an effective protect against downey mildew and botrytis. Think of willow in its environment, often rooted near water and having to counter humidity. It contains salicylic acid (as found in aspirin) which can overcome sap blockage issues in cold weather. Best to use new bark from willow whips as it contains more active ingredients. We harvest ours from the banks between the Kennet and Avon Canal and a tributary of the River Kennet, the Dundas. Once picked, we put the whips through a garden

shredder to make small chips. This is then simmered – never boiled as this would lose much of the salicylic acid – with water in a covered pan.

The effectiveness of these preparations is heavily dependent on the care of the practitioner in making them, the timing of harvesting each plant, their effective drying, and finally how they are stored before use.

- Biodynamic teas
- Compost tea (organic manure, seaweed, mollases)
- Horsetail
- Yarrow (sulphur)
- Chamomile (sulphur, potassium, calcium)
- Nettle (nitrogen)
- Oak bark (calcium, tannic acid)
- Willow bark (salicin)
- Dandelion (potassium, silicon)
- Valerian (warmth for frost)
- Bentonite clay (barrier to botrytis)
- Potassium bicarbonate (preventative against botrytis)
- Goose grass
- Seaweed
- Rapeseed
- Garlic

Horn manure BD 500

Horn Silica BD 501

Yarrow BD 502

Chamomile BD 503

Stinging Nettle BD 504

Oakbark BD 505

Dandelion BD 506

Valerian BD 507

Common Horsetail BD 508

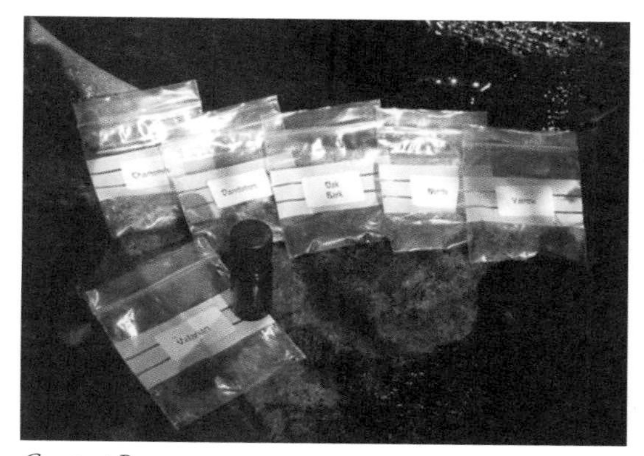

Compost Preps

Appendix 6 - Biodynamic Calendar

A YEAR IN THE LIFE OF A BIODYNAMIC VINEYARD

MONTH	DECEMBER	JANUARY ☽ ☍ ♄	FEBRUARY
SUMMARY	*The grapes have been harvested, most of the leaves have fallen from the vines leaving the bare canes. The energy of the plants has moved from above to below ground and the soil biome continues to work throughout the winter.*	*A quiet time, no growth above ground in the vineyard, too wet to work on the soil, a time for the farmer to rest, reflect, recuperate and plan the year ahead.*	*A quiet time, no growth above ground in the vineyard, too wet to work on the soil, yet a time of hope and beginning to look forward and see the first signs of life*
SUN CYCLE	**Winter Solstice 21st December** – the shortest day, the longest night. The point in the year with the fewest sun hours. A time of celebration and reflection.	The start of the year, a time of hope, resting and drawing inwards; the days are short, the nights are long. There is little change in daylight hours	**Imbolc (February 1st):** mid way between the winter solstice and the spring equinox. The increase in daylight hours is beginning to speed up
VINE STAGE	47 – 1 Dormancy / winter bud	1 Dormancy / winter bud	1 Dormancy / winter bud
VINE & LAND CARE	None	**Pruning** (using pruning paste and garlic maceration) **Hedges** cut/laid, **tree** planting, trellising/fencing **maintenance**	**Pruning** (using pruning paste and garlic maceration) **Hedges** cut/laid, **tree** planting, trellising/fencing **maintenance**
BD TREATMENTS & NATURAL PREPARATIONS	Usually too cold and wet to do soil treatments, although if the weather is suitable (>10°C) then an **additional BD500 / Compost Tea** soil spray might be beneficial. **Moon opp Saturn -** BD500 / Compost Tea, if poss	**Three Kings** preparation (if applied) **Moon opp Saturn -** Three Kings, if applied	None

MONTH	MARCH	APRIL	MAY
SUMMARY	*The beginning of spring, the first flush of flowers, the sap is beginning to rise and the early tree buds are bursting.*	*Mid spring, meadows of spring flowers, leaves are coming out, first butterflies, birds are nesting, a time of increasing energy*	*The height of spring moving towards summer; the lime-green of new leaves everywhere, vibrant and new, full of upward life energy*
SUN CYCLE	**Spring Equinox (March 21st) -** mid way between the shortest day and the longest day. Day and night of equal length. The increase in daylight hours is at its greatest.	The increase in daylight hours begins to slow a little, the clocks have changed giving more evening daylight hours	**Beltane (1st May) -** mid way between the Spring Equinox and Summer Solstice; the increase in daylight hours is beginning to slow as the year builds towards summer
VINE STAGE	1 **Dormancy / winter bud** (possibly beginning to swell)	**2-6 Bud Burst**	**7-18 Early shoot and leaf growth**
VINE & LAND CARE	**Pruning** (if sap rising no need for pruning paste) First **mowing** of alleys (allows for summer flowers and breaks up prunings); **strim** under vines (not too short); **scythe** wildflower areas and **lay hay** under weaker vines.	**Finish pruning** (if sap rising no need for pruning paste) **Spread BD compost** throughout vineyard if not done previous autumn **Vine planting**	**Bud rubbing** after last frost
BD TREATMENTS & NATURAL PREPARATIONS	1st **BD500 Horn Manure & BD compost tea** soil spray (descending moon, just before full moon if poss) Dig up **BD500 Horn Manure**, bury **BD501 Horn Silica**, create **BD compost** from manure **Moon opp Saturn -** BD500 / Compost Tea, if poss	2nd **BD500 Horn Manure & BD compost tea** soil spray (descending moon, just before full moon if poss); If buds have burst and there is potential for frost then **Valerian, Horsetail and Willow tea spray**; 1st health spray of **Dandelion, Horsetail, Willow & Nettle Tea plus Comfrey concentrate in organic whey/water** 50/50 mix (spray in morning) **Pick Dandelions** for teas and dry for BD506; **Excavate Yarrow** (bladder), **Chamomile** (intestine), **Oak Bark** (skull) and **Dandelion** (mesentery); **Build BD Compost** for compost tea **Moon opp Saturn -** Choose most important spray	3rd **BD500 Horn Manure & BD compost tea** soil spray (descending moon, just before full moon if poss); If buds have burst and there is potential for frost then **Valerian, Horsetail and Willow tea spray**; 2nd health spray of **Dandelion, Horsetail, Willow & Nettle Tea plus Comfrey concentrate in organic whey/water** 50/50 mix (spray in morning) + dilute Copper/other if needed Fill bladder with **Yarrow** and hang in sun over summer **Moon opp Saturn -** Choose most important spray

MONTH	JUNE	JULY	AUGUST
SUMMARY	*The mid point of the year, all the buds have burst and the vine shoots and leaves are growing rapidly in the longer hours of sun. A time of strong upward growth.*	*The month of flowering, moving towards the height of summer, flowers are transforming to become the first fruit, while grain is ripening*	*The month of fruit set, the first chance to see the potential harvest, moving towards the end of summer, the vine begins to focus its energy into the fruit*
SUN CYCLE	**Summer Solstice (21st June) -** the number of daylight hours are at their peak in June, with little change for 2-3 weeks over the solstice. A time of celebration.	The daylight hours slowly begin to reduce after the summer solstice, although there are still long days and short nights	**Lammas (1st August)** – mid way between the Summer Solstice and the Autumn Equinox, the days are becoming noticeably shorter, while the temperatures remain high
VINE STAGE	**7-19 Early shoot and leaf growth, beginning of flowering**	**19-27 Flowering and fruit set**	**27-35 Berry formation, swelling and ripening of early varieties**
VINE & LAND CARE	**leaf stripping** (first 2-3 leaves around flowers?) **Strim under disease susceptible varieties** (eg. Pinot Noir) to reduce humidity if necessary	**leaf stripping** (first 2-3 leaves around flowers?) **strim under, and mow alleys between, disease susceptible varieties** (eg. Pinot Noir) to reduce humidity if necessary; **tucking in** of vigorous shoots if necessary	**tucking in** of vigorous shoots **mow alleys before installing bird netting** (early varieties);
BD TREATMENTS & NATURAL PREPARATIONS	4th **BD500 Horn Manure & BD compost tea** soil spray (descending moon, just before full moon if poss); 3rd health spray of **Chamomile, Horsetail, Willow & Nettle Tea plus Comfrey concentrate, in organic whey/water** 50/50 mix (morning spray) + dilute Copper/other if needed 1st **BD501 Horn Silica** spray later in month (asc. moon, just before new moon if poss.); **Pick and bury nettles** for BD504; **collect and dry Horsetail** for use later in year and next spring; **collect and dry Chamomile** **Moon opp Saturn -** Choose most important spray	4th health spray of **Yarrow, Horsetail, & Willow Tea plus Comfrey concentrate in organic whey/water** 50/50 mix (spray in morning) + dilute Copper/other if needed 2nd **BD501 Horn Silica** spray (asc. moon, just before new moon if poss.); **Collect and dry Yarrow flowers** for teas and BD502; **collect and dry Horsetail** for use later in year and next spring; **pick and infuse Valerian flowers** for BD 507 **Moon opp Saturn -** Choose most important spray	5th health spray of **Yarrow, Horsetail, & Willow Tea plus Comfrey concentrate in organic whey/water** 50/50 mix (spray in morning) + dilute Copper/other if needed 3rd **BD501 Horn Silica** spray (asc. moon, just before new moon if poss.); **Collect and dry Yarrow flowers** for teas and BD502; **excavate BD504 Nettle** prep; **collect Oak Bark** for BD505 and bury in skull; **Moon opp Saturn -** Choose most important spray

MONTH	SEPTEMBER	OCTOBER	NOVEMBER
SUMMARY	*The 1st month of ripening, the 1st potential harvest, a time of gratitude and celebration.*	*The 2nd month of ripening and harvest. The beginning of leaf fall for early varieties, later in the month.*	*the 3rd and final month of ripening and harvest. Leaf fall for mid/late varieties. Temperatures usually dropping significantly with rainfall increasing.*
SUN CYCLE	**Autumn Equinox (21st September) -** mid way between the longest day and the shortest day. Day and night of equal length. The decrease in daylight hours is at its greatest.	The decrease in daylight hours begins to slow a little after the equinox while the clocks change giving even fewer evening daylight hours	**Samhain (1st November)** – traditionally the festival marking the end of harvest, the end of growth, the beginning of the shift to winter
VINE STAGE	**34-38 Softening and ripening of berries**	**37-38 Ripening of berries**	**37- 47 Ripening of berries through to leaf fall**
VINE & LAND CARE	**tucking in** of vigorous shoots**; bird netting** as varieties ripen**; harvest** of early varieties **mow alleys** just before bird netting needs to be installed	**harvest** of mid varieties; **removing bird netting** from harvested vines	**harvest** of late varieties; **removing bird netting** from harvested vines **spread BD compost** throughout vineyard if weather allows (after all grapes harvested)
BD TREATMENTS & NATURAL PREPARATIONS	6th health spray of **Yarrow, Horsetail, & Willow Tea plus Comfrey concentrate in organic whey/water** 50/50 mix (spray in morning) for late varieties; 4th **BD501 Horn Silica** spray (asc. moon, just before new moon if poss.) **Excavate BD501 Horn Silica**; **bury BD500** Horn Manure, **BD502** Yarrow, **BD503** Chamomile & **BD506** Dandelion **Moon opp Saturn -** Choose most important spray or a specific **harvest /pied-de-cuve** day	**Moon opp Saturn -** specific **harvest /pied-de-cuve** day	5th **BD500 Horn Manure & BD compost tea** soil spray (descending moon, just before full moon if poss), after all leaves have fallen and before the temperatures drop too low >10^{0}C. **Moon opp Saturn -** BD500/compost tea or a specific **harvest /pied-de-cuve** day

Appendix 7

Wine Labelling Explained

The gold foil is made from aluminium. It is slotted over the wire hood by a foiling machine. The neck label is then wrapped around the bottom of the gold foil.

The muselet plaque, branded, in our colours, sits on the top of the cork. The wirehood is fitted on top of both and the bottom twisted underneath the lip of the neck of the bottle to secure both.

The glass bottle must be thick enough to withstand seven bars of pressure.

Neck label edged with gold foil, with a circle on one end of the strap showing the county of origin as Berkshire.

Front label is edged with gold foil. It shows prominently the name of the vineyard, that the wine is English in origin, the style of wine ie brut reserve, and the method of making, ie 2 fermentations. It follows rules as set down by the Wine Standards Authority standards.

Back label. It shows an OS map of 1900 with the address of the vineyard at Orpenham Farm in the centre of the section. Then the vintage, that it is English sparkling wine, a short description, who made the wine, the grape varieties, the disgorging date (both optional), and that it contains some sulphites. It shows the alcohol content and the volume. This information gives the consumer all he needs to know about the wine's provenance.

Appendix 8
Low Intervention Wine Making

Intervention

Character (+risk)

No SO_2 at press
Natural settling of juice
Spontaneous fermentation
V low or no SO_2
Non-innoc MLF
Oak/terracotta dominance
No fining or filtration
Yeast and sugar tirage
Movement by gravity wherever possible
Minimal or no dosage

Small amount of SO_2 at press
Natural settling
Pied de cuve – ambient yeast starter culture
Innoculated MLF
Low/Med levels of SO_2
Stainless steel and oak
No fining or filtration
Yeast, sugar and riddling aid tirage
Gravity and pump movements
Low dosage with some SO_2

Considerable SO_2 at press
Enzymatic settling
Innoculated with commercial yeast
Innoculated with MLF
Considerable use of SO2
100% stainless steel
Fining and filtration
Yeast, sugar, riddling aid, nutrient and tannin tirage
Movements by pump
Med/High dosage with considerable SO_2
and additives

Appendix 9 - Statistics from WineGB and Wine Standards - June 2022

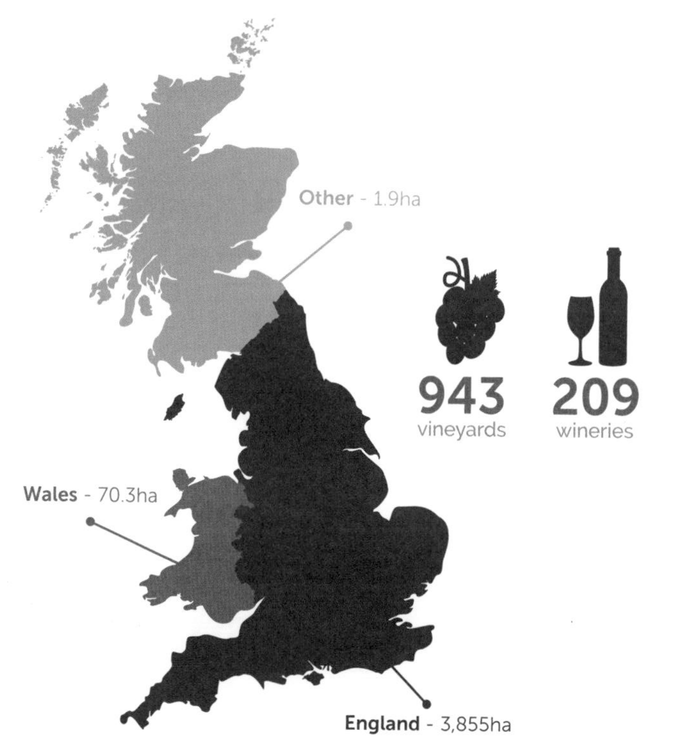

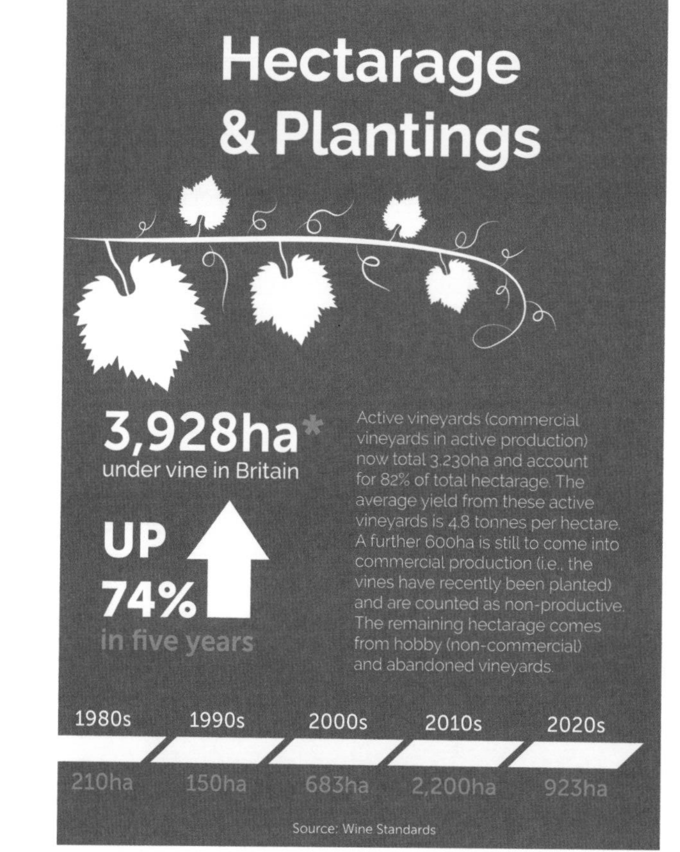

Planting in Detail

Most Planted Counties

England

County	(ha)
Kent (26% of total plantings)	1,033
West Sussex (15% of total plantings)	570
East Sussex (13% of total plantings)	493
Hampshire (10% of total plantings)	380
Essex (8% of total plantings)	325
Surrey (3% of total plantings)	127
Gloucestershire (2% of total plantings)	95
Devon (2% of total plantings)	92
Dorset (2% of total plantings)	90
Suffolk (2% of total plantings)	65

Wales

County	(ha)
Monmouth (0.6% of total plantings)	22
Carmarthenshire (0.3% of total plantings)	11
Vale of Glamorgan (0.2% of total plantings)	9
Powys (0.2% of total plantings)	8
Gwynedd (0.2% of total plantings)	7
Ceredigion (0.2% of total plantings)	6
Denbighshire (0.1% of total plantings)	2
Pembrokeshire (0.1% of total plantings)	2
Anglesey (0.04% of total plantings)	1.5
Conwy (0.03% of total plantings)	1.2

Most Planted Grape Varieties

Top ten variety	(ha)
Chardonnay (31% of total plantings)	1,228
Pinot Noir (29% of total plantings)	1,141
Pinot Meunier (9% of total plantings)	343
Bacchus (8% of total plantings)	298
Seyval Blanc (3% of total plantings)	122
Solaris	95
Reichensteiner	72
Pinot Noir Précoce	66
Rondo	61
Pinot Gris	58

Chardonnay, Pinot Noir and Pinot Meunier account for around 70% of total plantings. Hybrids, including PiWis, now make up 10% of all plantings. New entrants to the GB wine industry account for 55% of new plantings.

In the top 10 most planted counties, Gloucestershire has climbed one spot from last year's report and now sits ahead of Devon. However, half of new plantings are now taking place outside of the top 10 counties. In 2022, there were an additional 40ha planted in Kent, 23ha in Essex, 20ha in Hertfordshire, 19ha in Herefordshire, 18ha in Dorset and an additional 6ha planted in West Sussex, Norfolk, Worcestershire, East Yorkshire, and Northamptonshire.

Source: Wine Standards

4

Planting in Detail

Regional Plantings

Most planted varieties:

East

(Bedfordshire, Cambridgeshire, Essex, Hertfordshire, Norfolk, and Suffolk)

Variety	(ha)
Pinot Noir	145
Bacchus	99
Chardonnay	92
Pinot Meunier	26
Pinot Blanc	16

Thames & Chilterns

(Berkshire, Buckinghamshire, and Oxfordshire)

Variety	(ha)
Pinot Noir	33
Chardonnay	32
Pinot Meunier	9
Bacchus	9
Ortega	3

Midlands & North

(Cheshire, Cumbria, Derbyshire, East Yorkshire, Lancashire, Leicestershire, Lincolnshire, Northamptonshire, Nottinghamshire, North Yorkshire, Rutland, Shropshire, Souwth Yorkshire, Staffordshire, Warwickshire, West Midlands, West Yorkshire, and Scotland)

Variety	(ha)
Solaris	25
Seyval Blanc	21
Pinot Noir	19
Rondo	17
Bacchus	10

South East

(East Sussex, Kent, London, Surrey, and West Sussex)

Variety	(ha)
Chardonnay	811
Pinot Noir	724
Pinot Meunier	231
Bacchus	137
Solaris	42

Wales

(All counties in Wales)

Variety	(ha)
Pinot Noir	11
Seyval Blanc	11
Chardonnay	11
Solaris	7
Rondo	6

Wessex

(Dorset, Hampshire, Isle of Wight, and Wiltshire)

Variety	(ha)
Chardonnay	163
Pinot Noir	115
Pinot Meunier	50
Bacchus	11
Seyval Blanc	6

West

(Cornwall, Devon, Gloucestershire, Herefordshire, Isles of Scilly, Somerset, and Worcestershire)

Variety	(ha)
Pinot Noir	103
Chardonnay	63
Seyval Blanc	43
Bacchus	32
Pinor Meunier	29

Source: Wine Standards - All figures rounded up to the nearest whole number

While Pinot Noir features in the top 5 most planted varieties in all seven regions, there is more variation in plantings of Chardonnay and Pinot Meunier. In the Midlands and North of England, Chardonnay and Pinot Meunier do not feature in the top 5 list, yet in the South East and Wessex they come first and third respectively. Popular Vitis vinifera crossing Bacchus is listed in the top 5 for all English regions, with significant plantings in the East and South East. Seyval Blanc, a hybrid variety, shows more regional variation in planting, with a significant presence in the West, Midlands, and North of England, and in Wales.

Fun facts: There is Cabernet Franc planted in North Yorkshire and Nottinghamshire; Cabernet Sauvignon in Herefordshire and Devon; Chenin Blanc in Kent and Carmarthenshire; and Merlot in Powys, East Sussex and Cornwall. Kent has the most Albariño. Suffolk is the hotspot for Riesling, while East Sussex is home to the highest number of Gewurztraminer vines. There are now over 17ha of Sauvignon Blanc planted in the UK, 8ha of PiWi grape Divico, and 4ha of Gamay.

5

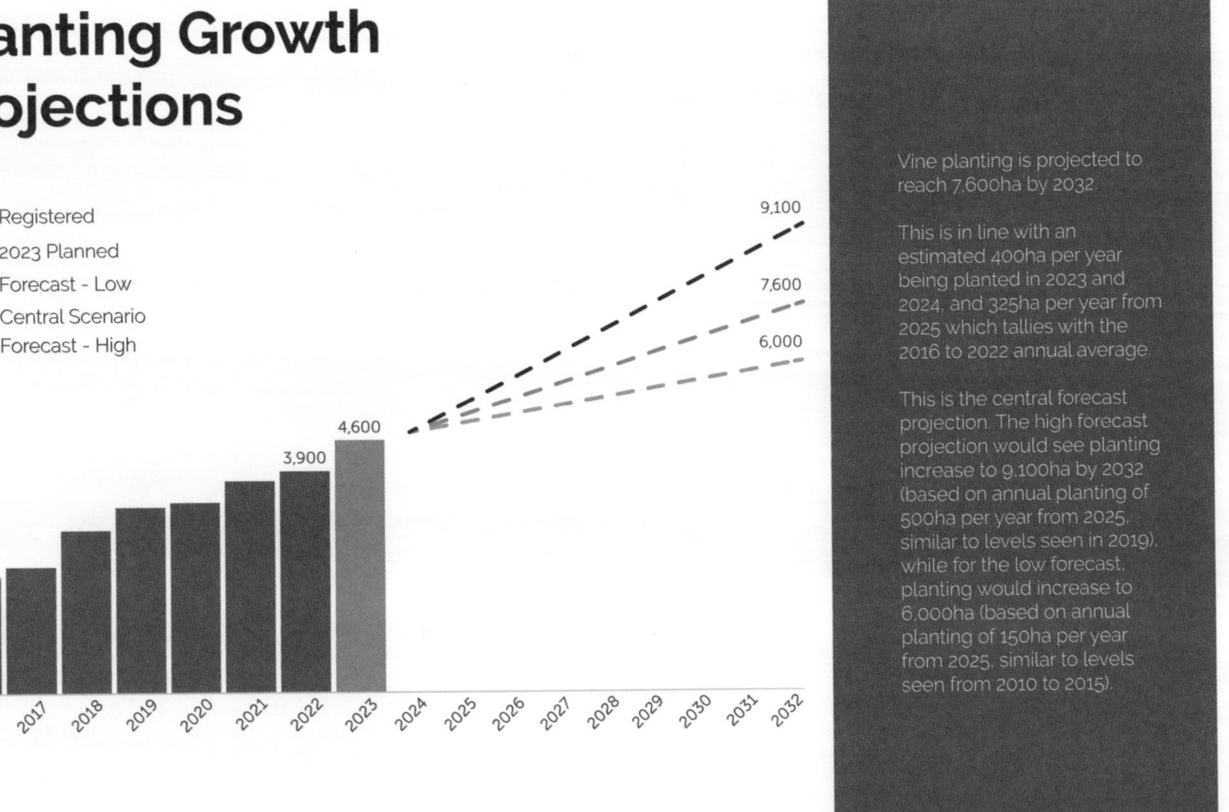
Planting Growth Projections
Registered
2023 Planned
Forecast - Low
Central Scenario
Forecast - High
9,100
7,600
6,000
4,600
3,900
2015
2016
2017
2018
2019
2020
2021
2022
2023
2024
2025
2026
2027
2028
2029
2030
2031
2032
Vine planting is projected to reach 7,600ha by 2032.
This is in line with an estimated 400ha per year being planted in 2023 and 2024, and 325ha per year from 2025 which tallies with the 2016 to 2022 annual average.
This is the central forecast projection. The high forecast projection would see planting increase to 9,100ha by 2032 (based on annual planting of 500ha per year from 2025, similar to levels seen in 2019), while for the low forecast, planting would increase to 6,000ha (based on annual planting of 150ha per year from 2025, similar to levels seen from 2010 to 2015).
6
Source: Figures cited are WineGB estimates based on historical Wine Standards data and figures obtained from the WineGB Industry Survey.

Production in Detail

Volume Production

	Bottles (m)	Sparkling	Still
2017	5.3	68%	32%
2018	13.1	69%	31%
2019	10.5	72%	28%
2020	8.8	64%	36%
2021	9.0	68%	32%
2022	12.2	68%	32%

Production figures from Wine Standards for 2022 represent a 36% year-on-year increase, driven by an increase in hectarage in active production and 20% increase in yields following the poor weather in 2021.

8.3m bottles of sparkling wine and 3.9m

Production Breakdown (2022)

Sparkling (Method)	
Traditional Method	**93%**
Charmat/Tank Method	**3%**
Carbonation	**3%**
Other	**<0.5%**

Sparkling (Style)	
White	**78%**
Rosé	**22%**

Still (Style)	
White	**62%**
Rosé	**21%**
Red	**16%**
Other	**1%**

12.2m bottles is equivalent to 91,323hl. 62% of the wine produced was made from vines/grapes owned by the winery, the highest percentage ever recorded, according to Wine Standards. The remaining volume was produced using grapes bought under contract.

Sales & Distribution

Sales by volume

2022 – 8m bottles

Sales distribution by style

5.6m sparkling (69%)
2.4m still (31%)

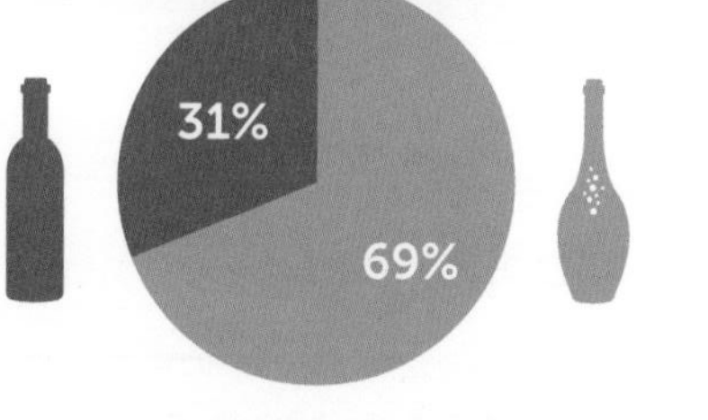

Distribution channels

Cellar door: direct-to-consumer	**19%**
Winery/vineyard website sales: direct-to-consumer	**11%**
UK on-trade (including sales via wholesalers)	**22%**
UK off-trade (national accounts)	**31%**
UK off-trade (independent retail)	**10%**
Export	**7%**

- This year's WineGB Industry Survey captured 91% of the total production (124 producers and growers), making it our most reliable and robust survey to date.
- The biggest 25 producers account for 83% of total sales.
- Direct-to-consumer sales total 30%, while the off-trade represents a 41% share.
- As the effects of Covid-19 wear off, purchasing habits have changed. This can be seen in the reduction of direct-to-consumer sales and growth in on- and off-trade accounts. Exports have risen to 7% of total sales, and with a record nine producers exhibiting on the WineGB ProWein stand in 2023, this looks set to continue.

Sales & Distribution

This is a new piece of data that shows the considerable variation in distribution channel according to producer size. The UK on- and off-trade remain the key markets for those that produce over 80,000 bottles, while cellar door sales are most significant for micro and small producers, making up to 12,000 bottles a year.

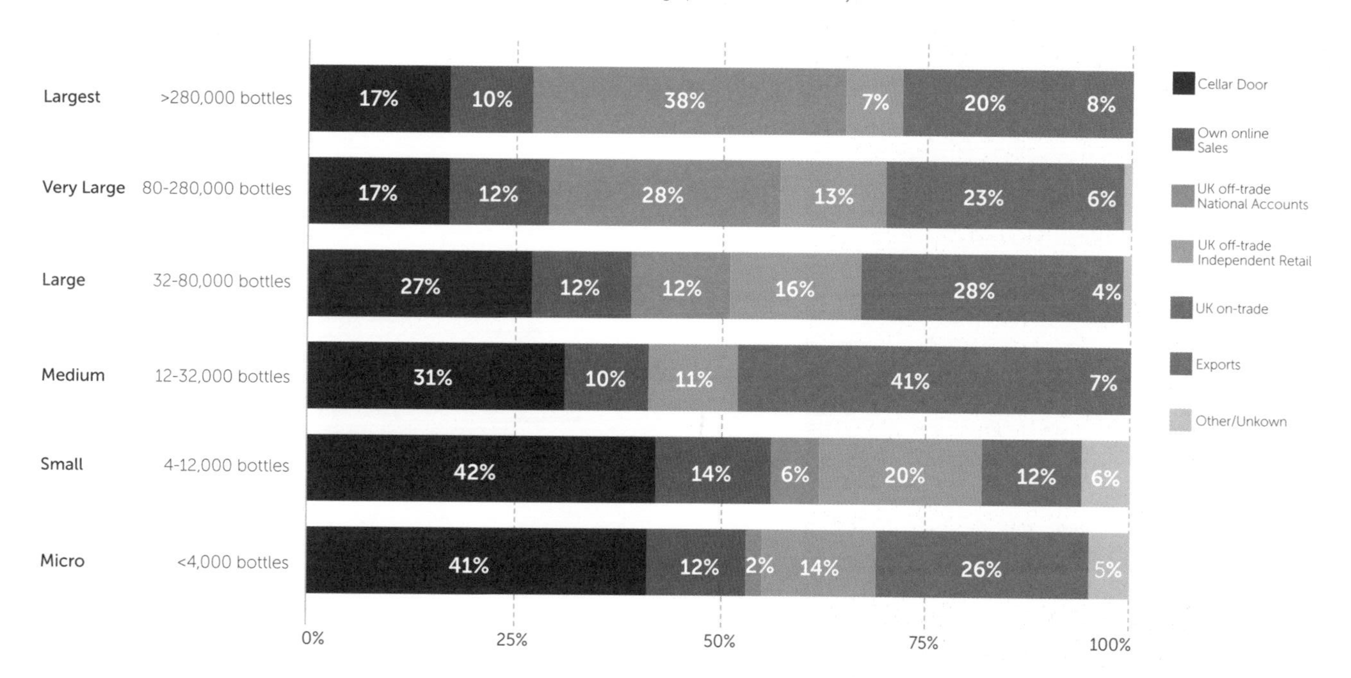

10

Wine Tourism

Tourist income as % of total revenue - by size

Tourism income % - by WineGB region

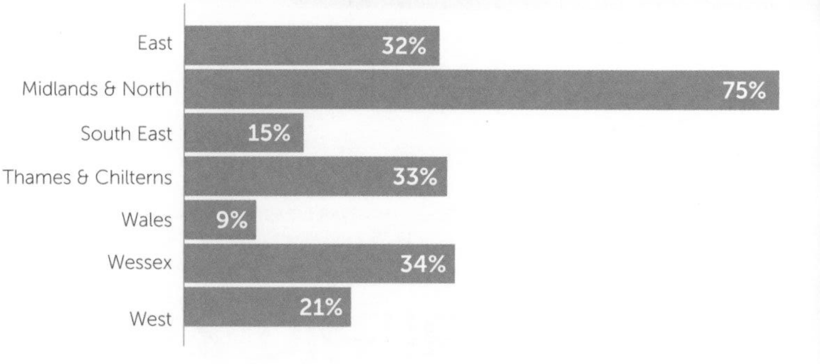

Income from wine tourism* averages at 24% of total revenue. Small to medium producers (up to 32,000 bottle production) are more reliant on direct-to-consumer sales and a physical sales presence (cellar door, café, events), particularly producers located in the Midlands and North of England.

Over 80% of producers offering wine tourism already have a cellar door presence and do tours and tastings. Only a third currently offer other tourism services (café, accommodation, events, etc) but many are planning to implement next year.

Based on figures from the WineGB Tourism Survey, conducted earlier this year, 2022 visitor numbers to GB vineyards and wineries rose 17% compared to 2021. However, the growth was uneven: 67% of survey respondents saw growth, others remained static. Half of those surveyed are expecting a greater than 20% increase in their visitor numbers over the next five years.

- Highest increases in the Midlands & North, Wessex and South East
- Inbound currently accounts for <10% of all visitors

Source: WineGB Industry Survey and WineGB Tourism Survey.

*Wine tourism income defined as: tour & tasting ticket sales; cellar door sales including merchandise; food and drink, accommodation, on-site events.

11

Appendix 10

Further Reading and Reference

Books

Dirt to Soil, Gabe Brown, 2018
English Pastoral, James Rebanks, 2020
English Wine, Oz Clarke, 2022
Wines of Great Britain, Ed Dallimore, 2022
Vines in a Cold Climate, Henry Jeffreys, 2023
A Biodynamic Manual, Pierre Masson, 2014
Biodynamic, Organic, and Natural Winemaking, Britt and Per Karlsson, 2018
Regenerative Viticulture, Jamie Goode, 2022
The Maria Thun Biodynamic Calendar, annual
Sunlight into Wine, Smart and Robinson, 2001 (out of print)
The Winegrowers' Handbook, Belinda Kemp & Emma Rice, 2012
From Vines to Wines, Jeff Cox, 1999
Vine Varieties, Clones and Rootstocks for UK Vineyards, S Skelton, 3rd edition, 2024

Websites

biodynamic.org.uk – British Biodynamic Association
englishwine.com – Stephen Skelton
greatbritishwine.com – John Mobbs
henryjefferys.substack.com – Henry Jefferys
jancisrobinson.com – Jancis Robinson
matthewjukes.com – Matthew Jukes

regenerativevitivulture.org – Regenerative Viticulture Foundation
sixatmospheres.substack.com – Tom Hewson
soilassociation.org – Soil Association
timatkin.com –Tim Atkin
thehivewine.com – Justin Howard - Sneyd
threewinemen.co.uk – Oz Clarke, Tim Atkin, Olly Smith
vineyardmagazine.co.uk
winegb.co.uk – trade body for UK wine industry
wineanorak.com – Jamie Goode

Interesting English vineyard and wine merchant websites
alburyvineyard.com
carteblanchewines.com
domainehugo.co.uk
grapebritannia.com
harrowandhope.com
hawkinsbros.co.uk
limeburnvineyard.co.uk
littlewadenvineyard.co.uk
tillingham.com
windingwoodvineyard.co.uk

Wine podcasts
Wine Blast with Susie and Peter
The jancisrobinson.com podcast
UK Wine Show
A Glass With – Olly Smith

English and Welsh organic and biodynamic vineyards and wineries (This list covers most commercial producers, either certified or practising, but it is not exhaustive)
Albury, Surrey
Ancre Hill, Monmouthshire
Bardfield Vineyard, Essex
Black Mountain Vineyard, Hereford
Bow-In-The-Cloud, Wiltshire
Davenport Vineyards, East Sussex & Kent
Domaine Hugo, Wiltshire
Flint Vineyard, Norfolk

Forty Hall Vineyard, Enfield
Hamstreet Wines, Kent
Harbourne Vineyard, Kent
Harrow and Hope, Buckinghamshire
Hebron, Pembrokeshire
Kingscote Vineyard, Sussex
Little Waddon Vineyard, Dorset
Limeburn Hill Vineyard, Bristol
Marden Organic Vineyard, Kent
Matt Gregory, Nottinghamshire
Mountain People, Monmouth
Offbeat Winery, Wiltshire
Oxney Organic Vineyard, East Sussex
Pebblebed Vineyard, Devon
Quoins Organic Vineyard, Somerset
Secret Valley Vineyard, Somerset
Sedlescombe Organic Vineyard, East Sussex
Sophie Evans Wines, Kent
Terlingham Vineyard, Kent
Tillingham, East Sussex
Trevibban Mill, Cornwall
Vineyard Farms, Sussex
Walgate & Co, Sussex
Westwell, Kent
Winding Wood Vineyard, West Berks

Acknowledgements

Thanks to the following

Robin Snowden, Limeburn Hill Vineyard, for permission to use his biodynamic calendar material.

WineGB and Wine Standards for use of their trade statistics.

Daniel Ham of Offbeat Wines for use of his diagram showing guidelines on low intervention wine making.

Caroline Finlayson for her wonderful artwork which adorns our tasting room, depicting the various stages of wine making. It is reproduced in the chapter on the art of wine making.

Charlie Stebbings for use of one of his fine snaps for our front cover. He knows how to use the camera on an iPhone which puts us all to shame. But he is a professional!

Ed Dallimore and Mick Berkout for kind use of their photographs.

To both Polly and Simon for so assiduously reading the proofs and correcting me on history, technical matters and a few prejudices held unfairly by me as a small wine producer.

To all my friends who allowed me to use photographs of them working (without make-up) in the vineyard.